T0282504

This book provides an introduction to lattice models of polymers. This is an important topic both in the theory of critical phenomena and in the modelling of polymers.

The first two chapters introduce the basic theory of random, directed and self avoiding walks. The next two chapters develop and expand this theory to explore the self avoiding walk in both two and three dimensions. Following chapters describe polymers near a surface, dense polymers, self interacting polymers and branched polymers. The book closes with discussions of some geometrical and topological properties of polymers, and of self avoiding surfaces on a lattice. The volume combines results from rigorous, analytical and numerical work to give a coherent picture of the properties of lattice models of polymers.

This book will be valuable for graduate students and researchers working in statistical mechanics, theoretical physics and polymer physics. It will also be of interest to those working in applied mathematics and theoretical chemistry.

CAMBRIDGE LECTURE NOTES IN PHYSICS 11
General Editors: P. Goddard, J. Yeomans

Lattice Models of Polymers

CAMBRIDGE LECTURE NOTES IN PHYSICS

Lattice Models of Polymers

CARLO VANDERZANDE

Limburgs Universitair Centrum, Belgium

CAMBRIDGE
UNIVERSITY PRESS

CAMBRIDGE UNIVERSITY PRESS
Cambridge, New York, Melbourne, Madrid, Cape Town, Singapore, São Paulo

Cambridge University Press
The Edinburgh Building, Cambridge CB2 8RU, UK

Published in the United States of America by Cambridge University Press, New York

www.cambridge.org
Information on this title: www.cambridge.org/9780521559935

First published 1998

A catalogue record for this publication is available from the British Library

Library of Congress Cataloguing in Publication data

Vanderzande, Carlo, 1958–
Lattice models of polymers / Carlo Vanderzande.
p. cm. – (Cambridge lecture notes in physics; 11)
Includes bibliographical references and index.
ISBN 0 521 55993 6
1. Polymers – Statistical methods. 2. Lattice dynamics.
3. Critical phenomena (Physics) I. Title. II. Series.
QC173.4.P65V36 1998
530.4′13 – dc21 97–32153 CIP

ISBN 978-0-521-55993-5 paperback

Transferred to digital printing 2008

To all my children

Contents

Preface and acknowledgements

Ideas from the theory of critical phenomena have been of great importance in the modelling of polymers ever since the Nobel prize winner P. G. de Gennes showed (in 1972) how the two subjects can be connected. In the 25 years that have passed since then, almost every major development in the understanding of criticality has led to parallel progress in the study of polymer models. We can think of the renormalisation group, the introduction of ideas from fractal geometry, conformal invariance .... As a result of all this work, the equilibrium behaviour of a polymer in a diluted regime is by now very well understood. That's why I considered the time ripe to write an overview of this field of research.

There already exist excellent books on the statistical mechanics of polymers and it may therefore be important to say a few words about the 'niche' in which this book should be placed. I have put the emphasis on models on a lattice and have therefore said very little about models, and methods to treat them, which work in the continuum. For completeness, important results obtained in the continuum are mentioned, but it would take much more space (and expertise on the part of the author) to treat them in all detail. Moreover, they have been very well described, for example in [65]. This book also deals almost exclusively with the very dilute regime in which we can study one isolated polymer and can neglect the influence of any other polymers which may be present. Within this area I have tried to give an overview of the work that was done using exact, analytical and numerical methods. Since the subject of this book is related to so many other themes in statistical mechanics it was necessary to make a further selection of topics. For example, the exact determination of some critical exponents for polymers in two dimensions relies on ideas of the theory of critical phenomena such as the Coulomb gas method, conformal invariance and exactly solvable models. Again, it would be impossible within the limits of this book to discuss all these techniques in any detail. I have therefore limited myself to a short presentation of the main concepts and ideas of these subjects. In a way they are treated as black boxes out of which we take the results relevant for polymers. I apologise to the reader who is left feeling unsatisfied after such a treatment.

The book is aimed at a reader who has already had a good undergraduate course in statistical mechanics, and who has some general knowledge of the theory of critical phenomena, for example at the level of [30]. My aim has been to write for such a reader an introduction to the study of lattice models of polymers. I have included an extensive list of references from which it is possible to delve further into the subject. This list does not pretend to be complete and I must apologise to these collegues who feel that their work has not been well treated here.

~

My Italian friends Attilio L. Stella and Flavio Seno have carefully read and made many useful comments on several chapters of this book. S. G. Whittington and A. Friedman of the Institute of Mathematics and its Applications in Minneapolis, USA, invited me to take part in the workshop on 'Geometry and topology in polymer physics', which has been very helpful in writing chapter 10. Murray Batchelor, Chris Soteros and Buks Janse van Rensburg answered particular questions I had during the writing of this book. Buks (and his coworkers) gave me permission to reprint figure 10.8 and figure 10.9 from one of his papers, while Mehran Kardar and Yakov Kantor gave permission to reprint some figures from their work on polyampholytes (figures 8.10 and 8.11). Rodney J. Baxter allowed me to reprint figures from his work as figure 3.5 and figure 3.6. I have used data from the thesis of Antonio Trovato to make tables 2.3 and 2.4. Enzo Orlandini lent me his nice set of pictures of self avoiding surfaces, one of which was used for figure 11.3. Geert-Jan Bex helped me in solving all problems that had to do with computers, and Conny Wijnants made all the drawings in this book. Hildegard was a big support (as she always is) during the writing of this book. Thanks!, grazie!, bedankt! to each of them.

1

From polymers to random walks

1.1 The world of polymers

Polymers are long chain molecules consisting of a large number of units (the monomers), which are held together by chemical bonds [1]. These units may all be the same (in which case we speak of homopolymers) or may be different (heteropolymers).

Chemists spend most of their time developing polymers with specific chemical or physical properties. Such properties are often determined by the characteristics of the monomers and their mutual binding. In other words, they are determined on a *local* scale. In contrast, physicists work in the spirit of Richard Feynman [2] and "have a habit of taking the simplest example of any phenomenon and calling it 'physics', leaving the more complicated examples to become the concern of other fields." This attitude is taken to the extreme in the statistical mechanics of polymers, where one is interested mainly in universal properties, i.e. those properties that depend only on the fact that the polymer is a long linear molecule, and are determined by 'large scale quantities' such as the quality of the solvent in which the polymer is immersed, the temperature, the presence of surfaces (on which a polymer can adsorb) and so on.

~

Having this in mind, we can introduce a description of polymers in terms of random and self avoiding walks. When we look at the polymer on a microscopic scale we remember from our chemistry courses that one of the binding angles between successive monomers is essentially fixed (like the well known 105° angle between the two H–O bonds in a molecule of water), leaving one rotational degree of freedom (figure 1.1) for the chemical bond.

Adding a large number of monomers to form a long polymer, one can understand that the correlation between bond directions

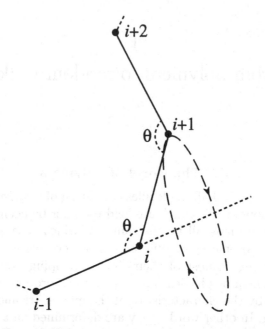

Fig. 1.1. Chemical bonds in a polymer. Monomers are shown as small spheres (i, $i+1$ etc.) and chemical bonds by straight lines. One angle (θ) between consecutive chemical bonds is fixed, leaving one rotational degree of freedom.

present on the monomer level will decay on larger length scales, until above a certain persistence length l_p, the directions of bonds become essentially uncorrelated. This persistence length can in principle be calculated from microscopic properties of the monomers. The idea of persistence length can be defined more precisely [3], but for us it is sufficient to work with the above intuitive idea to introduce a first model for a polymer chain.

1.2 The simplest model of a polymer

The independence of bond directions when one considers a polymer on length scales above the persistence length leads to the simplest model for an isolated polymer†. In this description, the

---

† In this book, we will limit ourselves to a discussion of polymers in the very dilute regime.

polymer is modelled as a *random walk* [4]. Random walks have a long history in mathematics and physics and the literature on them is huge. Here we will only review their properties as far as they are relevant to the description of polymers. This will serve as a warming up for the study of more realistic polymer models.

~

To be more specific, consider the hypercubic lattice $\mathbb{Z}^d$ in d dimensions. A random walker goes at each step with equal probability $1/(2d)$ from a site $\vec{x}$ to any of its nearest neighbour sites. Figure 1.2 shows an example of a long random walk.

After N steps there are $c_N = (2d)^N$ distinct random walks which each have the same probability of occurrence. We consider this set of random walks to be the set of allowed configurations for a polymer of N 'monomers' (elements of a discrete phase space). Time averaged properties of the polymer are then calculated as ensemble averages over this set of random walks. Before giving examples of such averages some remarks have to be made. Firstly, instead of the hypercubic lattice we could have considered a walk on any other regular lattice or in the continuum. This is not supposed to change any universal quantities. Secondly, it is important to point out that we refer to N as the number of monomers in the polymer, whereas strictly speaking it refers to a group of real monomers which has a length of the order of the persistence length.

Let us now calculate some simple polymer properties in our model. Firstly, the entropy S_N of the one-polymer system is simply given by

$$S_N = \log c_N = N \log 2d \qquad (1.1)$$

Secondly, let $\vec{R}_N$ be the end-to-end vector of the walk. What is the averaged squared length of $\vec{R}_N$? Clearly, $\vec{R}_N$ can be written as the sum of the (unit) vectors $\vec{a}_i$ which form the walk (which we will denote as W). In an obvious notation

$$\vec{R}_N = \sum_{i \in W} \vec{a}_i$$

Thus,

$$R_N^2 \equiv \langle \vec{R}_N \cdot \vec{R}_N \rangle_0 = \sum_{i,j \in W} \langle \vec{a}_i \cdot \vec{a}_j \rangle_0 \qquad (1.2)$$

Fig. 1.2. A random walk of 30,000 steps on a square lattice.

We will always denote by $\langle\ldots\rangle_0$ an average over the set of equally weighted random walks of fixed length (canonical average). In a random walk different steps are uncorrelated so that we get for (1.2)

$$R_N^2 = \sum_{i,j \in W} \delta_{i,j} = N \qquad (1.3)$$

It is common to introduce an exponent ν defined through the asymptotic behaviour of R_N^2

$$R_N^2 \sim N^{2\nu} \qquad (1.4)$$

So, in our simple random walk model, $\nu = 1/2$, independently of d. Such exponents which do not depend on space dimension are reminiscent of the mean field exponents in the theory of critical phenomena. Indeed, as we will explain below, the random walk model can be seen as a mean field approximation to a more realistic polymer model.

A special subset of the set of N-step random walks consists of those walks which after N steps have returned to their starting point. One can think of these as representing ring polymers. How many of them are there? Let's for simplicity calculate this quantity in one dimension. A necessary and sufficient condition is that the walk makes an equal number of steps to the left and to the right. Because the steps to the right can be performed at arbitrary times during the walk, the number q_N of walkers that return to the origin after N (even) time steps is

$$q_N = \binom{N}{N/2}$$

which using Stirling's formula becomes for large N

$$q_N \simeq 2^N N^{-1/2}$$

This result can be extended to hypercubic lattices in d dimensions with the result

$$q_N \simeq (2d)^N N^{-d/2}$$

Note that c_N and q_N grow exponentially, and with the same exponent. The corrections to this exponential term are often power laws and they are used to define critical exponents for polymers. In general one assumes that for arbitrary polymer models c_N and

q_N behave for 'large' N as

$$c_N \simeq \mu^N N^{\gamma-1} \tag{1.5}$$

$$q_N \simeq \mu^N N^{\alpha-3} \tag{1.6}$$

For random walks we thus obtain the exponent values $\gamma = 1$, $\alpha = 3 - d/2$ while the connective constant $\mu = 2d$.

$\sim$

In a grand canonical description of an isolated polymer, N is no longer fixed and its distribution is determined by a monomer fugacity z. The grand canonical partition function is then

$$\mathcal{Z} = \sum_{N=0}^{\infty} c_N z^N = \sum_{N=0}^{\infty} (2dz)^N = \frac{1}{1 - 2zd}$$

The last equality of course only holds for $z < z_c = 1/(2d)$. As a consequence, the average number of monomers diverges at z_c. Indeed, we have

$$\langle\langle N \rangle\rangle_0 = \frac{1}{\mathcal{Z}} \sum_{N=0}^{\infty} N c_N z^N = z \frac{\partial}{\partial z} \log \mathcal{Z} = \frac{2dz}{1 - 2zd}$$

Here $\langle\langle \ldots \rangle\rangle_0$ is used to denote grand canonical averages. Since in this book we are mainly interested in the critical behaviour of polymers, we write the above result as

$$\langle\langle N \rangle\rangle_0 \sim (z_c - z)^{-1}$$

Similarly one finds

$$R^2 \equiv \langle\langle R_N^2 \rangle\rangle_0 = \frac{1}{\mathcal{Z}} \sum_{N=0}^{\infty} R_N^2 c_N z^N$$

$$= \frac{2zd}{1 - 2zd} \sim (z_c - z)^{-1} \tag{1.7}$$

This ends our little tour of the properties of the random walk model. Despite its simplicity it has several merits. Firstly, there are situations in polymer physics in which a description through a random walk model is completely correct. Examples are polymers in $d \geq 4$, a polymer in a melt (a topic not discussed in this book) or a polymer at the so called θ-point in $d \geq 3$ (chapter 8). Secondly, the elementary calculations described above hint at a similarity between polymer physics and critical phenomena, a relation which will turn out to be very important.

Finally, we remark that one can easily introduce persistency in a random walk in order to model a polymer on length scales below the persistence length. Such models (see for example [5]) cannot in general be solved exactly, but have the property that for large N, relations like (1.4) still hold with $\nu = 1/2$, as might be expected on the basis of universality (no long range properties are changed; in the language of the renormalisation group persistency is said to be irrelevant).

1.3 Into the continuum

Although in this book we will mainly study models on a lattice, we will also have to discuss some properties of continuum models for polymers, since they are commonly used in, for example, the discussion of polymers in a random environment (a simple model of which we will discuss in the next section). Let's therefore investigate a continuum version of our random walk model. The usual place to start is from the master equation for a lattice random walk. Let $P(\vec{x}, t)$ be the probability that the random walker is in $\vec{x} \in (a\mathbb{Z})^d$ at time t (with an initial condition $P(\vec{x}, t_0) = \delta_{\vec{x}, \vec{x}_0}$), and let $\vec{e}_i$, $i = 1, \ldots, d$ be an orthonormal set. Finally we denote by a the lattice constant (which so far we have taken to be 1 for simplicity), and by τ the unit of time (i.e. time is discretised). Then, for a walker which makes a jump to a nearest neighbour site at each time step, it is obvious that

$$P(\vec{x}, t + \tau) = \frac{1}{2d} \sum_{i=1}^{d} \left(P(\vec{x} - a\vec{e}_i, t) + P(\vec{x} + a\vec{e}_i, t) \right)$$

To get a continuum representation of the random walk we now send a and τ to zero. Expanding P in a Taylor series gives

$$P(\vec{x}, t) + \tau \frac{\partial P}{\partial t} + \ldots = \frac{1}{2d} \sum_{i=1}^{d} \left(2P(\vec{x}, t) + a^2 \frac{\partial^2 P}{\partial x_i^2} + \ldots \right)$$

If we now assume that

$$\lim_{a, \tau \to 0} \frac{a^2}{2d\tau} = D \tag{1.8}$$

we arrive, after taking limits, at the familiar diffusion equation

$$\frac{\partial P}{\partial t} = D \Delta P \tag{1.9}$$

an interface between regions of up spins and down spins in an Ising ferromagnet. When the temperature is sufficiently low, this interface contains few overhangs, and is therefore directed when considered on a sufficiently large length scale.

~

The directed walk model is trivial to solve. To see this one has to realise that in the time direction the walker advances with constant speed, whereas in the perpendicular direction the motion is just that of a random walk. In general we can define two ν-exponents. In an obvious notation

$$\langle R_{\parallel,N}^2 \rangle_D \sim N^{2\nu_\parallel}$$
$$\langle R_{\perp,N}^2 \rangle_D \sim N^{2\nu_\perp} \qquad (1.14)$$

where $\langle \dots \rangle_D$ denotes an average over the set of all directed walks. So for the simple directed walk we have in general dimension $\nu_\parallel = 1$, $\nu_\perp = 1/2$.

~

In the last 10 years there has been a very great interest in the directed walk in a *random environment* [6]. In such a model one associates an energy E_b with each bond b (or site) of the lattice. The energies of different bonds are taken as independent, but identically distributed, random variables. Let us denote the distribution of bond energies by $P(E_b)$. In this way each directed walk W gets an energy E_W which is taken as the sum of the energies of the bonds which are traversed by W. Instead of giving an equal weight to all walks, we now assign weights according to the Boltzmann distribution, i.e. the weight of a given path is proportional to $\exp(-\beta E_W)$ where β is the inverse temperature. One is then interested in the properties of the partition sum

$$Z_N^D(\beta) = \sum_W \exp(-\beta E_W) \qquad (1.15)$$

where the sum is over all N-step directed walks. The model we have introduced is usually referred to as the directed polymer in a random environment (DPRM). For the trivial case $\beta = 0$, we recover the simple directed walks discussed above. On the other hand, for $\beta \to \infty$, the partition sum is dominated by an optimal path, which has the minimum energy, E_0. The value of E_0 depends on the configuration of bond energies. Its fluctuation is used to define a new exponent ω

$$\overline{E_0^2} - \overline{E_0}^2 \sim N^{2\omega} \tag{1.16}$$

The overline represents the average over all configurations of bond energies. One of the main challenges of the DPRM problem is to determine ω, and to determine the exponent $\nu_\perp$ for the minimal energy path.

In principle, the DPRM model is of little relevance for polymer physics. There are however several reasons to discuss it here. Firstly, it shows how techniques and methods used in the study of polymers have been taken over in other fields of statistical mechanics. Secondly, the DPRM may be considered as a first step towards the understanding of the behaviour of real polymers in random media. The DPRM has also become a kind of paradigm in the study of random systems in statistical mechanics, since it is much simpler than for example a spin glass [7], yet its behaviour is highly nontrivial. Moreover the DPRM is encountered in several problems of high current interest such as interfaces in random magnets, growth phenomena [8], or flux lines in superconductors. We will not go into these connections, but refer the interested reader to some nice reviews on this subject [6, 8]. A final reason to discuss the DPRM in this book is didactical: we will use it to illustrate the use of the replica technique and a simple case of the Bethe *Ansatz*.

$\sim$

In order to study the model defined in (1.15), we will use three approaches; firstly we discuss some rigorous properties of the partition function, secondly we discuss a numerical approach and finally we discuss a continuum version of the model, which within some approximations can be solved in two dimensions. This will illustrate the general methodology which we will also use in the next chapters where we will combine insights from rigorous, analytical and numerical work to come to an understanding of a given polymer model. This section by no means pretends to be a general overview of the work done on the DPRM.

$\sim$

To begin, we must make a remark about averaging. Notice that the partition sum (1.15) is a random variable. The corresponding free energy

$$F_N^D(\beta) = \log Z_N^D(\beta) \qquad (1.17)$$

is also random. For very large N we then proceed as follows. We break the directed walk into large pieces. The partition function $Z_N^D(\beta)$ can then be written as a product of partition functions for the pieces. The contributions of each of these to the partition sum will essentially be independent. Sure enough there will be some correlations, but if we are not too close to a critical point, these correlations will decay over a length that is small compared to the size of the pieces. In that case the free energy (1.17) will become a sum of random variables with short range correlations. The law of large numbers still applies to this situation and we therefore get

$$\lim_{N\to\infty} \frac{F_N^D(\beta)}{N} = \lim_{N\to\infty} \frac{\overline{F_N^D(\beta)}}{N} \qquad (1.18)$$

The relation (1.18), which we will assume to be valid, is also referred to as *self averaging*, since it implies that we don't have to perform the average when the walk is long enough. The average of the free energy is also referred to as the quenched average since we keep the randomness fixed in calculating the free energy. The 'opposite' case, in which the randomness is averaged in the partition function, is referred to as the annealed average. From the convexity of the log-function it follows that the annealed average gives an upper bound to the free energy (1.18)

$$F_N^D(\beta) \leq \log \overline{Z_N^D(\beta)}$$

It is easy to calculate this annealed average. Indeed we have

$$\overline{Z_N^D(\beta)} = \sum_W \overline{\exp\left(-\beta E_W\right)}$$

Since the energies are independent the average can be factorised. If we are on a $(d+1)$-dimensional hypercubic lattice (for the rest of this section, we isolate the time direction), there are $(2d)^N$ directed walks of N steps. Hence we get

$$\overline{Z_N^D(\beta)} = \overline{\exp\left(-\beta E_b\right)}^N (2d)^N \qquad (1.19)$$

From (1.19), we get for the annealed free energy density

$$\begin{aligned} f_a^D(\beta) &= \lim_{N\to\infty} \frac{1}{N} \log\left(\overline{Z_N^D(\beta)}\right) \\ &= \log\left(2d\right) + \log\left(\overline{\exp\left(-\beta E_b\right)}\right) \qquad (1.20) \end{aligned}$$

While this annealed free energy is an upper bound to the real free energy, it can be shown that at sufficiently high temperature it coincides with the exact free energy. The proof was first given by Cook and Derrida [9]. It uses known properties of the intersections of random walks. We refer the reader to the original reference where it is shown that there exists a temperature $T_2(d)$ such that for all higher temperatures the free energy is surely equal to the annealed one. On the other hand, the annealed free energy cannot be the exact one for all temperatures since when one calculates the entropy from (1.20), it is found to become negative below some temperature $T_1(d)$. (This lower temperature depends on the probability distribution $P(E_b)$.) We therefore conclude that there must exist some critical temperature $T_c(d)$, with $T_1(d) \leq T_c(d) \leq T_2(d)$, below which the free energy becomes different from the annealed one. For higher temperatures, we are in the annealed regime and the critical properties of the directed walk are those at $\beta = \infty$, i.e. the critical exponents are those when no randomness is present. For temperatures below T_c, we are in a new regime where the exponent $\nu_\perp$ takes on a new value. When we are in $d \leq 2$, the upper bound $T_2(d) = \infty$, so unfortunately, no conclusion can be made about the existence of a transition at finite temperature.

Therefore we have to turn to other approaches to find out whether there is a transition in $d \leq 2$. A first approach is numerical, and was performed by Derrida and Golinelli [10]. The idea is to set up an iterative scheme to calculate $Z_N^D(\beta, \vec{r})$ which is the partition function (1.15) restricted to those walks that end at $\vec{r}$. This partition function obeys the recursion relation

$$Z_{N+1}^D(\beta, \vec{r}) = \sum_{i=1}^{2d} \exp\left(-\beta E_{(\vec{r} \to \vec{r} + \vec{e}_i)}\right) Z_N^D(\beta, \vec{r} + \vec{e}_i)$$

where the $\vec{e}_i$ are the unit vectors perpendicular to the time direction, and $E_{(\vec{r} \to \vec{r} + \vec{e}_i)}$ is the energy along the bond going from $\vec{r}$ to $\vec{r} + \vec{e}_i$. This recursion relation can easily be iterated. Derrida and Golinelli consider this recursion for directed polymers which are constrained to a long strip of width W (with periodic boundary conditions). This is in fact a simple version of the transfer matrix for self avoiding walks which we will encounter in chapter 3. From the partition function, one can calculate the specific heat using standard thermodynamic relations. A phase transition should show up as a peak in the specific heat which grows when W is increased. In this way, the phase transition which was predicted in $d > 2$ was also found numerically. Numerical evidence was also obtained for a phase transition in $d = 2$. For the DPRM in $1 + 1$ dimensions the transfer matrix approach fails to find a transition. We will show below that in this case $T_c = \infty$, implying that the DPRM is always in the low temperature phase.

For the $(1 + 1)$-dimensional case it is possible to obtain $\widetilde{\nu}_\perp$ exactly. There are several ways to obtain this result. Here we will use a continuum approach, as was first done by Kardar [11] (the discussion below will be rather technical). A continuous version of the DPRM can be defined in the form of a path integral as follows

$$Z \equiv P_{\vec{x},t} \equiv \int_{0,0}^{\vec{x},t} \mathcal{D}\vec{y}(t) \exp\left(-\int_0^t \left[\frac{1}{4D}\dot{\vec{y}}^2(t') + \xi(\vec{y},t') \right] dt' \right) \quad (1.21)$$

where we have added a noise term $\xi(t)$ to the path integral (1.12). This term describes the random environment. The noise can be taken to be a Gaussian with mean 0 and variance σ^2. Furthermore it is uncorrelated in space and time

$$\overline{\xi(\vec{x},t)\xi(\vec{x}',t')} = \sigma^2 \delta(t - t')\delta(\vec{x} - \vec{x}') \quad (1.22)$$

From (1.21), we find that P obeys the partial differential equation

$$\frac{\partial P}{\partial t} = D\Delta P + \xi P$$

which is a Schrödinger equation (for imaginary time) of a particle in a random potential. We need to calculate the quenched average $\overline{\log P_{\vec{x},t}}$ of the free energy. Such a calculation, which always shows up in the statistical mechanics of random systems, is often performed using the replica trick. This technique uses the basic equality

$$\log x = \lim_{n \to 0} \frac{x^n - 1}{n}$$

and then interchanges the limit and the average over randomness

$$\overline{\log P_{\vec{x},t}} = \lim_{n \to 0} \frac{\overline{P_{\vec{x},t}^n} - 1}{n} \quad (1.23)$$

The replica trick was first introduced in the study of spin glasses [12]. Although there is no good understanding of why it works, it is quite commonly used and in general gives results which are in excellent agreement with those obtained from other approaches, or from numerical work.

We now use the replica trick to calculate the free energy of the DPRM in $1 + 1$ dimension. It is not too difficult to calculate the average of $P_{x,t}^n$. In fact we will calculate something more general

$$\overline{P(x_1, \ldots, x_n; t)} = \int_{0,0}^{x_1,t} \mathcal{D}y_1(t) \ldots \int_{0,0}^{x_n,t} \mathcal{D}y_n(t)$$

$$\exp\left[-\frac{1}{4D}\sum_{i=1}^{n}\int_0^t \dot{y}_i{}^2 dt_i\right] \overline{\exp\left[-\sum_{i=1}^{n}\int_0^t \xi(y_i, t_i) dt_i\right]}$$

The remaining average can be calculated by a cumulant expansion. Using (1.22) and assuming a Gaussian distribution for ξ we then obtain

$$\overline{P(x_1, \ldots, x_n; t)} = \int_{0,0}^{x,t} \mathcal{D}y_1(t) \ldots \int_{0,0}^{x,t} \mathcal{D}y_n(t)$$

$$\exp\left[-\frac{1}{4D}\sum_{i=1}^{n}\int_0^t \dot{y}_i{}^2 dt_i + \frac{1}{2}t\sigma^2\sum_{i\neq j}\delta(y_i - y_j) + \frac{1}{2}n\sigma^2 t\right]$$

It is now more convenient to go back to a partial differential equation for $P(x_1, \ldots, x_n; t)$

$$\frac{\partial P}{\partial t} = D\sum_{i=1}^{n}\frac{\partial^2 P}{\partial x_i^2} + \frac{1}{2}\sigma^2\left[n + \sum_{i\neq j}\delta(x_i - x_j)\right]P \qquad (1.24)$$

This problem is equivalent to the quantum mechanics of n particles in one dimension, interacting with a δ-potential. The solution can be written as an expansion in the eigenfunctions of the operator which occurs on the right hand side of (1.24). For long times t (i.e. for long 'polymers'), the solution will be determined by the largest eigenvalue of this operator (the corresponding eigenfunction $P_{0,n}$ will be called the 'ground state').

Let us first look at the case $n = 2$. In that case, the problem reduces to that of finding the ground state for a particle in a δ-potential. This elementary problem is solved in introductory courses on quantum mechanics, and the ground state solution $P_{0,n=2}$ is

$$P_{0,n=2}(x_1, x_2) \sim \exp-\left[\frac{\sigma^2}{8D}|x_1 - x_2|\right]$$

with ground state energy $E_{0,n=2} = (\sigma^4/8D) + \sigma^2$. For general n the ground state is found by making the so called Bethe *Ansatz* [13] (an introduction to the Bethe *Ansatz* can be found in [14]) in which the n-particle ground state is obtained as a product of the ground states for two-particle systems. More specifically, we have for this case

$$P_{0,n}(x_1, \ldots, x_n) \sim \exp - \left[\frac{\sigma^2}{8D} \sum_{i<j} |x_i - x_j| \right] \tag{1.25}$$

This kind of *Ansatz* for the solution of an n-particle problem once we know the solution in the two-particle case will be encountered in several situations later in this book. In this case, the form of the solution is rather simple and it therefore illustrates the method, which in the more complicated cases encountered later in the book will not be discussed in full detail.

The ground state energy of the n-particle problem is then found by inserting (1.25) in (1.24). The result is

$$E_{0,n} = n \left[\frac{\sigma^2}{2} - \frac{\sigma^4}{48D} \right] + \frac{\sigma^4}{48D} n^3 \tag{1.26}$$

Putting everything together and taking the limit $n \to 0$ we find that for large t the free energy of the DPRM is given as

$$\overline{\log Z} = \left[\frac{\sigma^2}{2} - \frac{\sigma^4}{48D} \right] t \tag{1.27}$$

showing that indeed the free energy is extensive, as should be the case.

More interesting is the sample-to-sample fluctuation of $\log Z$. To obtain this, we extend the replica trick and rewrite $\overline{Z^n}$ in a cumulant expansion

$$\overline{Z^n} = \overline{\exp n \log Z} = \exp \sum_{j=1}^{\infty} \frac{n^j}{j!} C_j(\log Z)$$

where C_j is the j-th cumulant. The first cumulant is the average and in this way we recover (1.27). As can be seen from (1.26), the only other non zero cumulant for large t is the third one

$$C_3(\log Z) \sim t$$

We can therefore conclude that the fluctuations in $\overline{\log Z}$ grow as $t^{1/3}$. This result is independent of σ which is the parameter that plays the role of inverse temperature. In particular it holds at zero temperature where the free energy coincides with the energy E_0 of the optimal path. Hence, we conclude that $\omega = 1/3$. As a final step in finding $\nu_\perp$, we invoke a scaling argument to determine a relation between ω and $\nu_\perp$. We therefore look at the minimal energy path which dominates the free energy at very low temperatures. Suppose then that the directed walk makes an extra excursion of size $x \sim t^{\nu_\perp}$ from this optimal, minimal energy path. We ask ourselves: how much will this change the free energy? The change can be estimated by the change in elastic energy

$$\int_0^t ds \, s^{2\nu_\perp - 2} \sim t^{2\nu_\perp - 1}$$

This leads to the scaling relation

$$2\nu_\perp = \omega + 1 \tag{1.28}$$

so that finally we obtain for the DPRM in $1 + 1$ dimension $\nu_\perp = 2/3$. Although many approximations have been used to obtain this result, it is generally believed to be exact. Indeed, the same result can be obtained from several other approaches, and it is in good agreement with numerical results.

We conclude by linking the results found in the continuum approach with those obtained on the lattice. In the first approach, when $\sigma = 0$ we get n decoupled free particles, and this then leads to random walk behaviour. But as soon as $\sigma \neq 0$, the δ-potential has one bound state and it is this fact that leads to a $\nu_\perp = 2/3$ for all T in $d = 1$. In $d \geq 2$, a weak potential has no bound states. Therefore, for small β, we recover the random walk behaviour. There is a critical value of σ (or β) above which the potential has a bound state and this is the quantum mechanical analogue of the phase transition which was predicted from the numerical work and from the exact bounds discussed at the beginning of this section.

In summary, the DPRM in $d + 1$ dimension is always in a new, low temperature phase when $d = 1$. In this phase, $\nu_\perp = 2/3$. In higher dimensions there is a high temperature phase where $\nu_\perp = 1/2$. At lower temperatures there is a transition to another phase. There exist only numerical estimates and educated guesses for $\nu_\perp$ in higher d. We refer the reader to the reviews mentioned at the beginning of this section for further information about the

DPRM. For us, it is time to go back to models more closely related to real polymers.

2
Excluded volume and the self avoiding walk

2.1 Self avoiding walks

In the discussion in the previous chapter we neglected the fact that polymers are almost always immersed in a solvent [1]. A good solvent is defined as a solvent in which it is energetically more favourable for the monomers of the polymer to be surrounded by molecules of the solvent than by other monomers. As a consequence, one can imagine that there exists round each monomer a region (the excluded volume) in which the chance of finding another monomer is very small. This will lead to a more open, more expanded structure for the polymer than if the excluded volume effects were absent.

The most popular model to describe this effect is the self avoiding walk [15]. Here one considers only the subset of random walks which never visit the same site again. An example is given in figure 2.1. When one compares this figure with that of a random walk, the excluded volume effect is obvious.

Thus, the equilibrium properties of a polymer with excluded volume effects are studied by making averages over the set of all N-step self avoiding walks (SAW) (we will encounter a 'continuum version' of the SAW model in chapter 4). All energy effects are taken into account by limiting the set of allowed configurations to the self avoiding ones. For the moment all SAWs therefore have the same energy and thus when we calculate averages, we weight all configurations equally. Note that the self avoidance constraint doesn't come from the fact that no two monomers can be in the same place, as is often stated. One must remember that the 'monomers' we speak about are already a collection of real monomers and thus there is no problem in them being on the same lattice site.

The questions which will occupy us in this chapter are similar

19

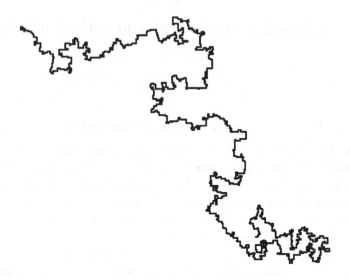

Fig. 2.1. A self avoiding walk of 2500 steps.

to those studied for random walks. For instance, how does the entropy depend on N? Are exponents such as ν, γ and α different from their random walk values? This can in fact be expected because self avoidance is a kind of long range effect, when one looks along the polymer chain. As we will discuss below, exponent values will indeed change, but determining their values or determining the growth constants like μ in equation (1.5) will turn out to be much more difficult than for random walks and few mathematically rigorous results exist. Nonetheless, we will see that the behaviour of SAWs, and thus polymers in good solvents, is well understood from a physicist's point of view.

2.2 The number of self avoiding walks

In this section, we discuss how to determine the number c_N of different N-step SAWs on a given lattice in $d > 1$.

There exist essentially two kinds of exact results on this quantity: these are exact numerical results for small N, and rigorous asymptotic results. Let us first discuss the latter, because they will yield insight on how to analyse the numerical, small N, data. It is easy to get bounds on the number c_N. Firstly, there must be fewer SAWs than random walks. Secondly, a lower bound is given by the number of directed walks. If we take walks of the type shown in figure 1.3a and work on a hypercubic lattice we get the following inequalities

$$d^N \leq c_N \leq (2d)^N \tag{2.1}$$

This result suggests that to leading order in N, c_N will grow exponentially. The way to prove this is by using a 'concatenation' argument. Let N, M be integers and consider the SAWs of $N + M$ steps. All of these can be cut in two parts creating an N-step SAW and an M-step SAW. On the other hand it is not obvious that concatenating any N-step SAW with an M-step SAW will lead to a walk that is self avoiding. We therefore have the inequality

$$c_{N+M} \leq c_N \, c_M \tag{2.2}$$

or

$$\log c_{N+M} \leq \log c_N + \log c_M \tag{2.3}$$

A series of numbers which obeys this property is called *subadditive* [16]. From (2.3) and the fact that c_N is bounded from above (2.1), it can be proved that

$$\lim_{N \to \infty} \frac{1}{N} \log c_N = \mu \tag{2.4}$$

exists (see e.g. [17]). We will refer to μ as a *connective constant*.†
From (2.2) it also follows that

$$c_N \geq \mu^N \qquad \forall N \geq 1 \tag{2.5}$$

The next question concerns the form of the finite N corrections to (2.4). A lot of hard mathematical work has been done on this problem. A famous result is Kesten's bound [18] which states that $\forall N \geq 2$

$$c_N \leq \mu^N \exp \left[A N^{2/(d+2)} \log N \right]$$

with A some constant. In fact in $d = 2$, the bound can be made somewhat sharper for large N [17]. This is how far exact mathematics can get. Further in this chapter we will give nonrigorous

† In this book we will denote a general connective constant by μ, while μ_d will be used for the case of the d-dimensional hypercubic lattice.

arguments from physics that show convincingly that, as already stated in (1.5),

$$c_N = \mu^N N^{\gamma-1} (B + \text{corrections}) \qquad (2.6)$$

where B and γ are constants.

For random walks we showed in the previous section that the connective constant is the same if one considers the subset of walks that return to the origin after N steps. Self avoiding walks can by definition never return to their starting point, but we can introduce the so called self avoiding polygon (SAP) or self avoiding ring which is a SAW whose $(N-1)$-th step brings it to a nearest neighbour site of the starting point. One can then imagine taking an N-th step which brings the walk back to the starting point thus forming a SAP. Let there be q_N of these N-step SAPs. Polygons that can be translated onto one another are not counted as distinct SAPs. We now give a concatenation argument that shows the existence of

$$\lim_{N\to\infty} \frac{1}{N} \log q_N \qquad (2.7)$$

in two dimensions. For SAPs such as those shown in figure 2.2 it is not obvious where the starting point is. We therefore introduce a *lexicographic ordering* on the vertices of the SAP. To achieve this, we first order the vertices of the polygon according to their first coordinate, say x. Vertices having the same x-coordinate are then ordered according to increasing y-values. Now, take two SAPs having respectively N and M vertices. Let the lexicographically largest point on the first polygon be at $\vec{y}_1$. Then we move the lexicographically smallest point of the second polygon to $\vec{y}_1 + \vec{e}_x - \vec{e}_y$. Then the two polygons can be concatenated as shown in the example of figure 2.2.

Since this procedure can be applied to all pairs of N- and M-step polygons we arrive at the result

$$q_{N+M} \geq q_N\, q_M$$

This is a result similar to (2.2) for SAWs and is sufficient to prove the existence of the limit in (2.7). With a slight generalisation this concatenation argument can be extended to higher dimensions [17]. Finally, a more detailed argument shows that the connective constants for SAWs and SAPs are equal [17]. In section 2.5 we

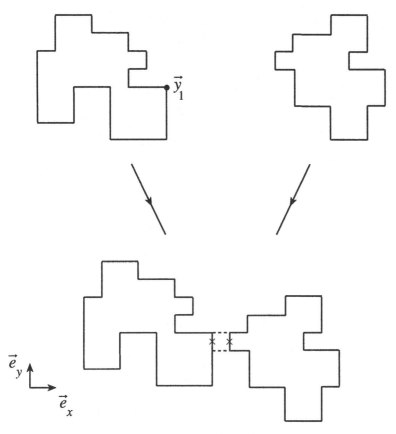

Fig. 2.2. Concatenation for polygons. The crossed edges are deleted and the polygons are joined along the dotted edges.

will understand this equality on a physical basis and we will argue that the full behaviour of q_N is of a form already given in (1.6)

$$q_N = \mu^N N^{3-\alpha} \left(\tilde{B} + \text{corrections} \right) \qquad (2.8)$$

where $\tilde{B}$ is a constant.

This is an appropriate place to briefly discuss the SAW in a grand canonical ensemble. Here we will encounter the first evidence of the many similarities which exist between polymer physics and critical phenomena, a topic which will be fully discussed in section 2.4. Consider therefore the situation in which the length of the SAW is not fixed but determined by a fugacity z. From (2.6) it follows that the grand canonical partition function will diverge when $z \to z_c$ as

$$\mathcal{Z}(z) = \sum_{N=1}^{\infty} c_N z^N \sim (z - \mu^{-1})^{-\gamma} \qquad (2.9)$$

At z_c several quantities will behave in a non-analytical manner. As the simplest example the average length of the SAW will diverge near $z_c = \mu^{-1}$ as

$$\langle\langle N \rangle\rangle = z \frac{\partial}{\partial z} \log \mathcal{Z}(z) \sim (z - \mu^{-1})^{-1} \qquad (2.10)$$

We will use $\langle\langle \ldots \rangle\rangle$ to denote averages over the set of all SAWs or SAPs. When $z > \mu^{-1}$ we are in a peculiar regime where the average length of the SAW is infinite. One refers to this as the *dense* phase of the polymer. We will study the properties of this phase further in chapter 7.

2.3 Exact enumerations

We now turn to a brief discussion of the method of exact enumeration of SAWs and/or SAPs which gives numerically exact values for c_N for 'small' N and which allows accurate estimates of μ, and of critical exponents. In the exact enumeration method one uses a computer to generate and to count all distinct SAWs or SAPs up to a given $N_{\max}$. Because of the exponential increase of c_N this counting can get quite time and memory consuming and clever algorithms have been devised which can now give results on the square lattice up to $N_{\max} = 51$ [19]. For the case of SAPs on the hexagonal lattice, it is possible to get up to $N_{\max} = 82$ [20], while on the cubic lattice the record for SAWs stands at $N_{\max} = 21$ [21].

As an example using the data from [19] we get from (2.5)

$$\mu_2 \leq 2.71815$$

Similarly, for polygons on the hexagonal lattice we get from the data in [20]

$$\mu_{\text{hex}} \geq 1.642$$

Besides just counting the number of SAWs, one can also calculate other quantities for all walks up to a given $N'_{\max}$. We can think of the average end-to-end distance, the average area inside a SAP, the number of 'contacts' (chapter 8) and so on. Such calculations take more time and have therefore only been performed for smaller walks and polygons.

A whole sophisticated machinery has been developed to analyse results from these exact enumerations and to get better estimates for μ than those following from (2.5). A full discussion of these methods is outside the scope of this book (for a reference, see [22]). We limit ourselves to a brief discussion of the ratio method and of the method of Padé approximants. Suppose we have a series of data d_N which for large N goes as

$$d_N \sim a^N N^b \qquad (2.11)$$

The ratio method consists of taking the logarithm of the ratios of consecutive terms of the series. These quantities should, for large N, be a linear function of $1/N$

$$\log \frac{d_{N+1}}{d_N} \simeq \log a + b(1/N)$$

A simple fit then gives estimates for a and b. A ratio analysis of the series for $\langle R_N^2 \rangle$ is even simpler since in that case $a = 1$. Ratio methods are simple but for short series often suffer from strong corrections to the form (2.11), for instance from effects of the lattice structure. A more advanced method uses Padé approximants. To understand how this method works, we will work with the example of the series c_N. Consider the function

$$h(z) = -\frac{d}{dz} \log \left(\sum_{N=1}^{\infty} c_N z^N \right)$$

From (2.11), it follows that for z approaching $1/a$, $h(z)$ will be given by $(b+1)/(z-1/a)$, plus corrections. Thus $1/a$ is a pole of order 1 of $h(z)$ and b can be obtained from the residue of $h(z)$ in $1/a$. That is one of the main reasons for considering the function $h(z)$. In contrast to the grand partition sum it is a meromorphic function. That is one of the requisites for a Padé approximation to work well.

The Padé method consists of finding estimates for the pole and residue of $h(z)$ from a partial knowledge of the coefficients c_N. From these coefficients we can obtain a polynomial approximation $(h_{N_{\max}}(z))$ to $h(z)$, correct up to order $N_{\max}$. To estimate $z_c = 1/a$ from the finite series we first write $h_{N_{\max}}(z)$ as the ratio of two

Excluded volume and the self avoiding walk

Table 2.1. *The connective constant on some lattices as determined using exact enumerations*

lattice	μ	reference
square	2.638159 ± 0.000002	[19]
hexagonal	1.847759 ± 0.000006	[20]
triangular	4.15080 ± 0.00003	[21]
hypercubic	4.68393 ± 0.00006	[21]
bodycentered cubic	6.53014 ± 0.00017	[21]

polynomials $P_K(z) = \sum_{k=0}^{K} a_k z^k$ and $Q_L(z) = \sum_{l=0}^{L} b_l z^l$

$$h_{N_{\mathrm{max}}}(z) = \frac{P_K(z)}{Q_L(z)} \qquad (2.12)$$

The coefficients a_k and b_l can uniquely be determined from (2.12) and the known series coefficients for any K and L such that $K + L = N_{\mathrm{max}}$. The polynomial $Q_L(z)$ has L (complex) zeros. The smallest real zero of these (when it exists) is then taken as an approximation for z_c. This estimate depends on K and L. Empirically it is known that the best results are found if one takes $K \approx L$ †. Finally one can get estimates of exponents like γ or ν from a calculation of the residue of $P_K(z)/Q_L(z)$ at the (approximate) pole of $h(z)$. This method works particularly well for the exponent ν since in that case we know that the pole of $h(z)$ is located at $z = 1$ exactly.

As a summary, we list in table 2.1 the current best estimates for μ as determined from exact enumerations (more sophisticated methods than those discussed here have been used to obtain the results in this table).

2.4 The exponent ν

In this paragraph we will discuss a very simple 'theory' for self avoiding walks which provides a formula for the exponent ν which is remarkably close to the best known values from numerics. This theory is known as the Flory theory [3].

---

† Some better arguments can be given for why this part of the Padé table gives the best results; see [22].

Consider a polymer consisting of N monomers. Our purpose is to calculate its average size $\langle R_N \rangle$. To do this, we will estimate the free energy $F(R)$ for fixed value R and then determine $\langle R_N \rangle$ ($\langle \ldots \rangle$ denotes an average over SAWs at fixed N) as that value of R which minimises F. Let us first estimate the entropy $S_N(R)$ of all N-monomer chains with a fixed R. We will approximate this entropy by its value for the random walk. If $c_N(R)$ is the number of random walks with radius R we can write

$$c_N(R) = c_N \ P_N(R)$$

where $P_N(R)$ is the probability that a random walker has reached a distance R after N steps. We have calculated this quantity in the continuum version of the random walk with the result (1.10). Thus

$$S_N(R) = \log c_N(R) = -B\frac{R^2}{N} + C$$

where B is a constant and C is a term which does not depend on R_N. Secondly, we need an expression for the energy $U_N(R)$ of the polymer. We can describe the effects of the good solvent by introducing a monomer–monomer repulsion which models the excluded volume effects. As a first and mean field assumption we take the corresponding energy density to be proportional to the density in the polymer ρ ($= N/R^d$) squared. After integrating this energy density over the polymer we then get

$$U_N(R) = A\frac{N^2}{R^d}$$

so that up to terms which do not depend on R we get

$$F_N(R) = A\frac{N^2}{R^d} + T \ B \ \frac{R^2}{N} \tag{2.13}$$

This free energy reaches a minimum when

$$\langle R_N \rangle = \tilde{A}N^\nu$$

with

$$\nu = \frac{3}{d+2} \tag{2.14}$$

This is the famous Flory formula which gives $\nu = 1, 3/4, 3/5$ and $1/2$ for dimensions from 1 to 4. The values in $d = 2$ and 4 are now thought to be exact results (see below), and the result in $d = 3$ is within 1% of the best numerical estimates. When

$d > 4$, $\nu < 1/2$ which implies that for large N, the term in $U_N(R)$ goes to zero and can therefore be neglected. So, above four dimensions, only the entropy term survives and we get the random walk result $\nu = 1/2$. This strongly suggest that $d = 4$ is the so called upper critical dimension [23] above which polymers in a good solvent behave like random walks. Intuitively, it is clear that as d increases there is less and less chance for the random walk to cross itself and it behaves therefore more and more like a SAW.

It is very remarkable that such a simple theory can work so well. Some attempts have been made to understand the success of the Flory theory [24, 25]. Several authors have tried to make similar theories to predict the ν-exponent for polymers in other situations, such as at the θ-point (chapter 8; [26]) or for a polymer in a random environment, or to calculate other exponents such as γ for SAWs [27]. None of these theories is as successful as the original Flory theory.

The result that $1/\nu$ is non-integral for SAWs implies that a polymer is a fractal. To see what this means we calculate the mass M of the polymer inside a region of size R. For a homopolymer, this mass is proportional to the number of monomers; we can thus write

$$M \sim N \sim R^{1/\nu} \qquad (2.15)$$

In general one 'defines' a fractal dimension D of an object through the relation [28]

$$M \sim R^D$$

(This is a kind of 'experimental' definition; for more mathematically correct definitions, see [28].) For ordinary, so called 'fat' objects, the fractal dimension $D = d$. For an object to be a *fractal* we should have $D < d$, implying that for large R the density goes to zero. For polymers in a good solvent, the Flory formula gives

$$D = \frac{1}{\nu} = \frac{d+2}{3}$$

In a fractal, because $D < d$ there are large holes, and if we look on larger length scales, the size of the holes also has to increase in order for the density to go to zero. One says that there are holes on all length scales.

Let us finally briefly return to random walks which are fractals in $d > 2$. A well known result for geometrical objects is the so called law of codimension additivity [28]. Let $\tilde{D} = d - D$ be the codimension of a (fractal) object. Then the law of codimension additivity states that generically the codimension of the intersection of two objects is the sum of their codimensions (think e.g. about two planes in $d = 3$; their intersection generically is a line). For the intersection of a random walk (which has $D = 2$) with itself this gives a dimension $D_{\text{int}} = d - 2(d - 2) = 4 - d$, implying that in $d > 4$ there is essentially no intersection between two random walks. This is in agreement with what we found in Flory theory.

We will come back to the exponent ν, and other exponents in the next chapter.

2.5 Relation with critical phenomena

In this section we will present a result, due to P.G. de Gennes [29], which relates the properties of SAWs to those of magnetic systems near their critical point. This relation, which was discovered around 1972, is of the greatest importance because it allows the application of all techniques and ideas from the theory of critical phenomena to polymer models.

~

To discuss this relation we will use a high temperature expansion. We assume that the reader is familiar with the high temperature expansion of the Ising model [30]. For further reference, however, we need to repeat some aspects of this technique. Consider therefore the partition function Z_{I} of the familiar Ising model

$$Z_{\text{I}} = \text{Tr} \prod_{\langle i,j \rangle} \exp\left(K s_i s_j\right)$$

where as usual $\langle i, j \rangle$ stands for pairs of nearest neighbour sites on a lattice, K is the reduced coupling constant, and the Ising spin s can be $+1$ or -1. If the lattice contains N_B edges we can rewrite Z_{I} as follows

$$\begin{aligned} Z_{\text{I}} &= \text{Tr} \prod_{\langle i,j \rangle} (\cosh K + s_i s_j \sinh K) \\ &= (\cosh K)^{N_B} \text{Tr} \prod_{\langle i,j \rangle} (1 + s_i s_j \tanh K) \end{aligned}$$

If we introduce $v = \tanh K$ and drop the trivial first factor, the

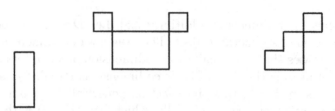

Fig. 2.3. Graph in the expansion of the Ising model partition function.

Ising model partition function can be rewritten as

$$Z_I = \text{Tr} \prod_{\langle i,j \rangle} (1 + v s_i s_j) \tag{2.16}$$

The high temperature expansion then consists of writing out the product in (2.16) in powers of v, and performing the trace, i.e. summing over spin values. All the (nonzero) terms in this expansion can be represented as a graph of a given weight. An example of such a graph in $d = 2$ is given in figure 2.3.

As can be seen, graphs consists of loops but these are not necessarily self avoiding. Furthermore a particular graph can contain several loops.

The idea is now to find another magnetic model for which the graphs in the high temperature expansion contain only one self avoiding loop. The way to achieve this is to work with a more general spin variable. Consider therefore an n-dimensional spin $\vec{s}$ of fixed length equal to $\sqrt{n}$, i.e. $\vec{s} \cdot \vec{s} = n$. This spin can be on the surface of a n-dimensional sphere. For the particular case $n = 1$ we recover the Ising spin. The next thing to do is to define a magnetic model involving these spins. This model will be referred to as the $O(n)$-model [31]. We could define such a model using a Hamiltonian, but since in the end it is the partition function that matters, we will, as is very common in the study of polymers, define the model through its partition function. An obvious generalisation of the Ising case (2.16) is to define the partition function Z_n of the $O(n)$-model as

$$Z_n = \text{Tr} \prod_{\langle i,j \rangle} (1 + v \vec{s}_i \vec{s}_j) \tag{2.17}$$

What we still have to discuss is what is meant by Tr in this case. Obviously it involves integrating all spin vectors over the surface of an n-dimensional sphere of radius $\sqrt{n}$. We will write this integration as $\int d\Omega$ and will include in the surface element $d\Omega$ a normalisation factor to ensure that

$$\int 1 \, d\Omega = 1 \tag{2.18}$$

It follows from rotational symmetry that for any component s_α, where $\alpha = 1, \ldots, n$, of the spin vector $\vec{s}$

$$\int s_\alpha^{2n+1} \, d\Omega = 0 \qquad \forall n \tag{2.19}$$

Furthermore, it is clear that

$$\int s_\alpha s_\beta \, d\Omega = \delta_{\alpha,\beta} \tag{2.20}$$

We can now start with the expansion in powers of v of the partition function (2.17). We will first discuss this expansion on the hexagonal lattice. This lattice has the advantage that there are at most three edges incident on any vertex. Any term in the expansion of the partition function can be visualised by a graph in which we indicate which particular edges, taken from the product in (2.17), contribute to that particular term. On the hexagonal lattice a given spin, say $\vec{s}_k$, can occur at most three times in any term of the expansion. In fact, because of (2.19) those terms in which a spin occurs once or three times are zero. Therefore we are left with graphs containing only loops (figure 2.4). Each of these loops is a SAP on the hexagonal lattice.

What is the weight of such a graph? Firstly, if the graph contains N bonds in total it must get a factor v^N. Secondly, we still have to integrate over spin directions. Because of the scalar products in (2.17), all spins in one loop are coupled, but spins in different loops are independent. Within one loop, (2.20) teaches us that all components have to be equal to get a nonzero contribution. We thus find that each loop contributes a factor n. So, if a particular allowed graph $\mathcal{G}$ contains L loops we get an extra weight factor n^L. Combining everything we finally get (on the hexagonal lattice!)

$$Z_n = \sum_{\mathcal{G}} v^N n^L \tag{2.21}$$

The $O(n)$ partition function in this elegant form was mostly popularised by B. Nienhuis [32], and because of the form of (2.21) the model is also referred to as the *loop gas*.

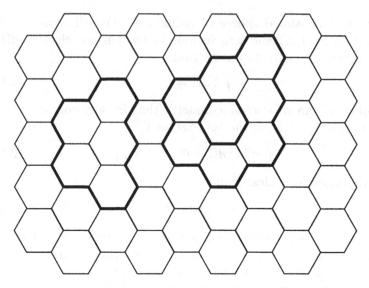

Fig. 2.4. Graph in the expansion of the partition function of the
$O(n)$-model on the hexagonal lattice.

Now to get graphs containing one SAP, we have to send $n \to 0$,
so that in (2.21) the leading contribution comes from graphs with
$L = 1$. More precisely, we can introduce, in the thermodynamic
limit, a free energy per site $f(n)$

$$f(n) = \lim_{N_s \to \infty} \frac{1}{N_s} \log Z_n$$

where N_s is the number of vertices of the lattice. When we take
$n \to 0$ we then get

$$\lim_{n \to 0} \frac{f(n)}{n} = \sum_{SAP} v^N \qquad (2.22)$$

where now the SAPs contributing to (2.22) all have the same lexicographically smallest site. This is just the grand canonical partition function for SAPs with a fugacity v.

Before further commenting on this result we discuss what hap-
pens on any other lattice. In that case four or more edges can be
incident on a given vertex, so that we have to calculate quantities
such as

$$\int (s_\alpha^4) d\Omega$$

We will not go into the details of calculating this quantity but
we just mention the fact that it will be proportional to n. In an
expansion of (2.17) on the square lattice, loops occurring in a
particular graph are no longer necessarily self avoiding. But loops
containing intersections get more powers of n than loops that don't
have intersections, and so when we finally send $n \to 0$ the result
(2.22) remains valid. Similar conclusions hold on lattices with a
higher coordination number such as the cubic lattice in $d = 3$.
Thus, while the result (2.21) is specific to the hexagonal lattice,
(2.22) is quite general.

So much for the relation between SAPs and the $O(n)$-model.
But what about SAWs? We will now show that they occur in the
calculation of spin–spin correlation functions. Take two spins $\vec{s}_k$
and $\vec{s}_l$ a distance r apart and calculate their correlation function

$$G_n(k,l) \equiv \langle \vec{s}_k \cdot \vec{s}_l \rangle = \frac{\text{Tr}\,(\vec{s}_k \cdot \vec{s}_l)\,\prod_{\langle i,j \rangle}(1 + v\vec{s}_i\vec{s}_j)}{Z_n} \quad (2.23)$$

We will again calculate this correlation function on the hexagonal
lattice, exactly as we did for the partition function. For a graph $\mathcal{G}_1$
to give a nonzero contribution to the correlation function, a self
avoiding path of bonds has to be present between sites k and l.
Besides this the graph can contain loops. A typical graph is shown
in figure 2.5. Thus, on the hexagonal lattice we have

$$G_n(k,l) = \frac{\sum_{\mathcal{G}_1} v^N n^{L+1}}{Z_n} \quad (2.24)$$

Now, we again send $n \to 0$ with the result

$$G_0(k,l) \equiv \lim_{n\to 0} G_n(k,l)/n = \sum_{\substack{\text{SAW} \\ k\to l}} v^N \quad (2.25)$$

Again, (2.25) can be shown to hold on any lattice, whereas (2.24)
is specific to the hexagonal lattice.

To get the grand partition function for SAWs we now sum
$G_0(k,l)$ over l. The resulting quantity is precisely the susceptibility χ_0 of the $O(n)$-model. We get

$$\chi_0 \equiv \sum_l G_0(k,l) = \sum_{\text{SAW}} v^N = \sum_N c_N v^N = \mathcal{Z} \quad (2.26)$$

which shows that the grand partition function of the SAW is equal
to the susceptibility of the $O(n = 0)$ model.

Excluded volume and the self avoiding walk

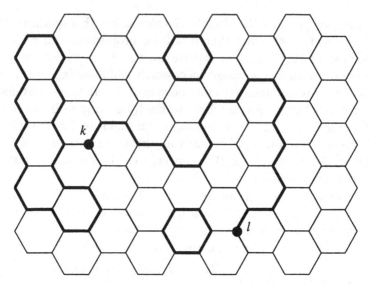

Fig. 2.5. Graph in the expansion of the spin–spin correlation function of the $O(n)$-model on the hexagonal lattice.

It is also of interest to investigate other correlation functions. One can for example associate an energy E_e with an edge e of the lattice. This energy is simply the scalar product of the spins at the two sites on which that edge is incident. Consider then the energy–energy correlation function $\hat{G}_n(e, d)$ defined as

$$\hat{G}_n(e, d) \equiv \langle E_e E_d \rangle \qquad (2.27)$$

Going through arguments similar to those just discussed one can see that this energy–energy correlation can be written as a sum over diagrams which consists of loops one of which has to go through both the edges d and e. When $n \to 0$ only this latter loop survives. Summing over all the edges d we get the specific heat C of the $O(n)$-model. We leave the derivation of these results as an exercise to the reader. In the polymer limit this specific heat is again given by a sum over self avoiding loops. Notice that in the spin–spin correlations one self avoiding path goes between the two sites k, l whereas in the energy–energy correlation function two such paths go between the vertices e and d. More generally, one can consider a correlation function $G_{n,L}(k, l)$ between vertices (or edges) k and l in which one sums over all graphs which contain

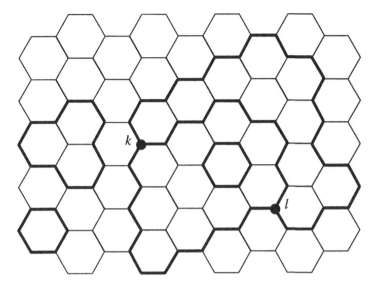

Fig. 2.6. Graph contributing to the watermelon correlation function with $L = 3$.

~

L self avoiding paths going from site k to site l, in the presence of other loops. Such correlation functions have been named 'watermelon' correlation functions (figure 2.6 shows an example with $L = 3$), a name introduced by B. Duplantier [33].

~

Let us now return to equation (2.9). It shows that the grand partition function for SAWs diverges at $1/\mu$ with the SAW exponent γ. On the other hand, one expects a ferromagnetic model like the $O(n)$-model to have a critical point (let's say at $v = v_c(n)$) where its thermodynamic properties behave in a non-analytical way. As an example the susceptibility of such a magnetic model diverges at $v_c(n)$ with an exponent $\gamma(n)$, and the free energy density f has a singularity characterised by an exponent $2 - \alpha(n)$. The exponent $\alpha(n)$ is in turn related to the correlation length exponent $\nu(n)$ using hyperscaling $\alpha(n) = 2 - \nu(n)d$. We remind the reader that these exponents can in turn be related to exponents that describe the decay of a correlation function [23]. For the watermelon correlation function $G_{n,L}(k,l)$ we introduce the exponent $x_L(n)$ which describes the decay of that correlation function at criticality

$$G_{n,L}(k,l) \sim |k - l|^{-2x_L(n)} \tag{2.28}$$

Then one has all the usual scaling relations such as

$$\gamma(n) = \nu(n)(d - 2x_1(n)) \tag{2.29}$$

$$\alpha(n) = \nu(n)(d - 2x_2(n)) \tag{2.30}$$

$$\nu(n) = 1/(d - x_2(n)) \tag{2.31}$$

(In the rest of this book exponents with an explicit n-dependence are those for the generic $O(n)$-model, whereas those for SAWs are denoted without writing $n = 0$.)

We now see that (2.22) and (2.26) tell us that the SAW exponents are just like ordinary critical exponents, in fact they are those of the (ferromagnetic) $O(n)$-model in the limit $n \to 0$. As a consequence, all methods used to study critical behaviour (such as the renormalisation group) can now be applied to the study of polymers. Moreover, combining (2.26) with (2.9) we see that μ is just the inverse of the critical temperature of the $O(n = 0)$-model. In the next two chapters we will exploit all these results and see that by now the behaviour of SAWs is rather well understood thanks to the equivalence with magnetism.

$$\sim$$

As a first corollary, we can derive some useful scaling relations for the number $c_N(r)$ of SAWs of N steps going between two points a distance r apart [15]. We start from the result (2.25) which we rewrite (with $r = |k - l|$) as

$$G_0(r) = \sum_N c_N(r) v^N \tag{2.32}$$

Near criticality a correlation function of a magnetic system has the usual scaling form,

$$G_0(r) = A r^{-2x_1} F(r/\xi) \tag{2.33}$$

where the correlation length ξ diverges near v_c as

$$\xi \sim |v - v_c|^{-\nu} \tag{2.34}$$

Combining equations (2.32)–(2.34) and performing an inverse Laplace transform (here one has to assume a Tauberian property; see [34]) we arrive at the following scaling form for $c_N(r)$, which should hold for large N:

$$c_N(r) \approx (v_c)^{-N} N^{\gamma - 1 - \nu d} H\left(\frac{r}{N^\nu}\right) \tag{2.35}$$

where H is a scaling function. We also used (2.29) in deriving this result. Now, notice that when r becomes very small a SAW just becomes a SAP. So in that limit $H(x) \sim x^{(\gamma-1)/\nu}$ in order to recover the usual form for q_N. Alternatively, we can rewrite (2.35) as

$$c_N(r) \approx q_N r^{(\gamma-1)/\nu} \hat{H}(\frac{r}{N^\nu}) \qquad (2.36)$$

in which $\hat{H}$ is another scaling function. Notice that for the case of random walks, equation (1.11) is indeed of the form (2.35) with $\nu = 1/2$ (we could introduce a probability $p_N(r) = c_N(r)/c_N$ that an N-step SAW reaches a distance r from its starting point).

$\sim$

As a second consequence of the relation between magnetic and polymer models, we mention that it is well known that ferromagnetic models have an upper critical dimension d_c above which their critical exponents take on mean field values ($\nu = 1/2$ and $\gamma = 1$). For the $O(n)$-model, $d_c = 4$, which for polymers implies that above four dimensions a SAW has mean field exponents, or equivalently has random walk exponents. This result can in fact also be derived rigorously without any use of the $O(n)$-model using a technique called the lace expansion [17, 35]. It is also the result which we already anticipated on the basis of Flory theory.

3

The SAW in $d = 2$

In the previous chapter we learned the importance of the $O(n)$-model for the study of polymers. In this chapter we will see how in two dimensions the critical behaviour of the $O(n)$-model has been determined exactly. The critical exponents of this model were first conjectured from renormalisation group arguments by Cardy and Hamber [36]. These conjectures were then confirmed by Nienhuis [37] using an approximate mapping onto the 'Coulomb gas'. Since the Coulomb gas can be renormalised exactly, this lent support to the belief that the exponents of Cardy and Hamber were indeed exact. In 1986, Baxter [38] succeeded in solving the $O(n)$-model exactly on the hexagonal lattice. His results, which were obtained in the thermodynamic limit, were later extended to finite systems by Batchelor and Blöte [39]. In more recent years an $O(n)$-model on the square lattice has received a lot of attention since it has a very rich critical behaviour.

We conclude this chapter with a discussion of the SAW on fractal lattices.

3.1 The Coulomb gas approach to the SAW in $d = 2$

In this section we discuss how the $O(n)$-model (2.17) can be related to the Coulomb gas. This relation holds for the $O(n)$-model on the hexagonal lattice. It will give exact results for exponents ν and γ which because of universality should also hold on other lattices. Furthermore an exact value for μ follows from this mapping. When we talk about an exact solution we must note that the result is derived by nonrigorous means but that nevertheless it is generally believed that the result is the exact one. All numerical calculations furthermore support these conjectured values.

A detailed discussion of the Coulomb gas and its critical be-

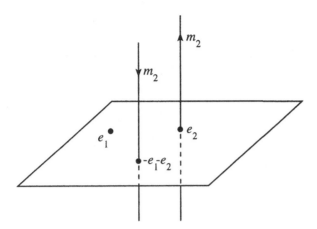

Fig. 3.1. A configuration of the Coulomb gas with three charges. The magnetic charge m_2 can be thought of as arising from an infinitely long wire with current m_2.

haviour is lengthy, and requires a lot of background information, and is therefore beyond the scope of this book. Here we will limit ourselves to a brief outline of the method (for a more general discussion, see [40]).

~

We understand by the Coulomb gas a gas of charged particles that interact electromagnetically. We will limit ourselves to the two-dimensional case which is most relevant for the study of critical phenomena. Besides electric charges the particles can also have magnetic charges (or one can think of the magnetic charges as representing the current per unit length in a infinitely long wire that is oriented perpendicular to the two-dimensional plane, figure 3.1). Let us define a lattice version of the Coulomb gas. We consider a lattice where each site can either be empty or contain a particle. Particles are distinguished according to their electric charge e and magnetic charge m. When both kind of charges are present, charge is quantised and can only take on integer values. The particles interact through Coulomb potentials which in $d = 2$ are logarithms (this is also the form of the potential between two infinitely long current-carrying wires). Furthermore, we remind the reader that in the presence of both electric and magnetic charges Maxwell's equations have a duality in which electric and magnetic fields can be interchanged. This duality requires electric and magnetic interaction strengths to be each other's inverse [41]. The magnetic coupling constant will be denoted by $g/2$. The number of particles with charges (e, m) is determined by a fugacity $z_{e,m}$. Finally we require total charge neutrality of the gas. The grand partition function Z_{CG} of the Coulomb gas is then given by

$$Z_{\text{CG}} = \text{Tr} \exp \mathcal{H}$$

where the reduced Hamiltonian $\mathcal{H}$ is

$$\mathcal{H} = \sum_{e,m} \log z_{e,m} \; + \; \sum_{k,l} (\frac{1}{2g} e_k e_l + \frac{g}{2} m_k m_l) u_e(k - l)$$

$$+ \; \sum_{k,l} i e_k m_l u_m(k - l) \qquad (3.1)$$

where e_k, m_k are the charges at the lattice site k. The first sum goes over all integer values of e and m while the second and the third sums go over all pairs of sites on the lattice (when $e_j = m_j = 0$ the site j contains no particle). The functions u_e and u_m are the electrostatic and magnetostatic potentials which are given as respectively the real and the imaginary parts of the complex logarithm. In summary, we see that the Coulomb gas has infinitely many couplings: g and all the fugacities $z_{e,m}$.

The critical behaviour of the Coulomb gas can be obtained from a renormalisation group (RG) study of this model. That means that one rescales the model with an (infinitesimal) factor b and sees how the coupling constants change as a function of b (in fact this is most easily done for a continuum version of the Coulomb gas). One can then look for the fixed points of the RG equations and critical exponents can, as usual, be determined by linearising the RG equations near the fixed point(s). It would lead us to far to go into the technicalities of this calculation, so we just summarise the results.

1. There is a line of fixed points along which the exponents change continuously. This line is given by the equations $z_{e,m} = 0 \; \forall e, m$ and corresponds physically to the 'vacuum' state. So independently of g the vacuum is critical with exponents varying as a function of g.

2. The RG equations can be calculated exactly near the vacuum up to second order in the fugacities. This is an extremely important result since it implies that the exponents which we obtain from linearisation are *exact*. Furthermore higher order contributions can be shown to be irrelevant.

3. After linearisation one finds the following set of renormalisation group eigenvalues

$$y_{e,m}(g) = 2 - \frac{e^2}{2g} - \frac{gm^2}{2} \qquad (3.2)$$

These exponents indicate that if we add a term with a small $z_{e,m}$ to the vacuum, this term will renormalise into $b^{y_{e,m}(g)} z_{e,m}$. Depending on the sign of $y_{e,m}$, this operator can be relevant or irrelevant.

In fact, the result (3.2) can be understood, without making any RG calculations, by looking at a two-point correlation function in the vacuum state. Let us therefore consider a system of just two particles (with charges e, m and $-e, -m$ (charge neutrality!)) a distance r apart. Their 'correlation function' $G_{e,m}(r)$ in the vacuum state is simply given by the exponential of their electrostatic energy

$$G_{e,m}(r) = \exp\left[-\left(\frac{1}{g}e^2 + gm^2 \right) \log r \right]$$

which can be rewritten as

$$G_{e,m}(r) = r^{-2\left[\frac{e^2}{2g} + \frac{gm^2}{2}\right]}$$

The decay of a correlation function as a function of distance is usually described by an exponent denoted $2x$ (see (2.28)) so in this case we have

$$x_{e,m}(g) = \frac{e^2}{2g} + \frac{gm^2}{2} \qquad (3.3)$$

If one remembers the well known relation between exponents $y = d - x$ one easily recovers in this way the result (3.2).

~

The results discussed so far rely on the assumption that the fugacities $z_{e,m}$ are symmetric under charge reversals, i.e. the assumption that $z_{e,m} = z_{-e,m} = \dots$. When this condition no longer holds the critical behaviour of the Coulomb gas becomes more difficult but also much richer. In that case also the coupling constant is renormalised and one has to use instead of a bare coupling g a renormalised coupling g_R. One can then find expressions similar to (3.3) which depend on g_R, but we will not go into these aspects.

Now that we have some knowledge of the critical behaviour of the Coulomb gas we can explain how one in general proceeds to determine the critical behaviour of another model (for example, the Potts model (chapter 6), or in the case of interest here, the $O(n)$-model) from that of the Coulomb gas. To achieve this goal one should be able to map the model under consideration onto the Coulomb gas. It is not *a priori* obvious that such a mapping is possible, but it turns out that it can be found for many of the popular models from statistical mechanics. In some cases such a mapping is exact, in other cases (like that of the $O(n)$-model) it requires further assumptions. The goal then is to relate the critical point of the model to a particular value of g. Equation (3.2) then gives a list of possible (relevant) exponents for that g-value. As a second step one has to relate specific correlations in the model under study to correlations of charges in the Coulomb gas.

To be more specific we will quote the results for the $O(n)$-model on the hexagonal lattice. Here one finds through a series of mappings [37] that $O(n)$ correlation functions have to be related to Coulomb gas correlations with fugacities that are not symmetric under charge reversal. As a consequence, one finds a relation between the number of spin components n and the renormalised coupling g_R which is

$$n = -2\cos(\pi g_R) \qquad (3.4)$$

with $1 < g_R < 2$. So for polymers $(n = 0)$ we have $g_R = 3/2$. We also have to relate the watermelon exponents $x_L(n)$ to Coulomb gas exponents $x_{e,m}(g_R)$. We only quote the result

$$x_L(n) = g_R \frac{L^2}{8} - \frac{(1 - g_R)^2}{2g_R} \qquad (3.5)$$

This leads, using (2.29), to the following exponents for the two-dimensional SAW problem

$$\nu = 3/4 \qquad \alpha = 1/2 \qquad \gamma = 43/32 \qquad (3.6)$$

Finally, we mention that on the hexagonal lattice it is possible to determine the critical point of the $O(n)$-model for general n. The result is

$$v_c(n) = \frac{1}{\left[2 + \sqrt{2 - n}\right]^{1/2}} \tag{3.7}$$

which for $n \to 0$ gives the result $\mu_{\text{hex}} = \sqrt{2 + \sqrt{2}}$.

This ends our little tour of the world of the Coulomb gas. When we discuss percolation phenomena in chapter 6 we will see that in that case too the Coulomb gas will allow the determination of a set of exact exponents.

3.2 The transfer matrix

Thanks to the Coulomb gas method we now have the predictions (3.6) for the exact values of exponents for SAWs in $d = 2$. Since these predictions involved some assumptions and approximations, it is necessary to find further support for them using other means. We have already discussed the method of exact enumerations, which gives exponent estimates very close to those found in the previous section. For two-dimensional systems there exists however another method which in general gives very accurate values for critical properties. This method uses the transfer matrix [30]. In fact, as we will see below, this method is nothing else but the exact enumeration of SAWs in a strip of the two-dimensional lattice.

The transfer matrix is well known in the study of spin models. Given the close relation between such models and SAWs it is natural to try to set up a transfer matrix approach to SAWs. Such a calculation was first performed by B. Derrida [42].

Let us consider a strip with periodic boundary conditions of width W on a square lattice. Our purpose is to calculate the spin–spin correlation function of the $O(n = 0)$-model between two points k and l which are situated a large distance from each other along the strip (see figure 3.2). We know that this correlation function is given in terms of the number $c_{W,N}(k, l)$ of N-step SAWs going between k and l

$$G_{0,W}(k, l) = \sum_N c_{W,N}(k, l) z^N \tag{3.8}$$

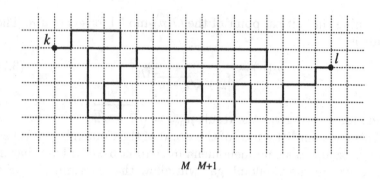

M   M+1

Fig. 3.2. A self avoiding walk in a strip.

Since a system in a strip cannot be critical we expect the correlation function $G_{0,W}(k,l)$ (where we have now added explicitly the W-dependence) to decrease exponentially with R, the distance between k and l

$$G_{0,W}(k,l) \sim \exp\left(-R/\xi_W(z)\right) \qquad (3.9)$$

Here $\xi_W(z)$ is a correlation length.

To set up a transfer matrix we first look at the edges of the strip between two consecutive columns (say M and $M+1$) and look for all possible configurations in which this set of edges can be. We will denote such a configuration by $\mathcal{V}$. The simplest configuration is that there is just one step of the SAW passing between these columns. But many more complicated configurations are possible. In order to be allowed, a configuration has to fulfil some conditions: there must be an odd number of steps of the SAW between the two columns, all but one of which are paired (see figure 3.2), and paired bonds cannot mutually cross. The allowed configurations can then be grouped using the symmetries (rotation and reflection) of the strip. In this way the number of allowed edge configurations V_L can be obtained (table 3.1 gives V_L for L up to 11).

We next define a function $H_M(\mathcal{V})$ which gives the (weighted) sum of all possible graphs (these don't have to be connected!) which end in configuration $\mathcal{V}$ in column M. The correlation function $G_{0,W}(k,l)$ which we want to calculate is $H_R(\mathcal{A})$ where $\mathcal{A}$ is the configuration with one horizontal step ending in site l. We

Table 3.1. *Results of phenomenological renormalisation for the SAW in a strip of width L, from [46]*

L	z_c	ν	V_L
2	0.347810385	0.6684732	1
3	0.365304779	0.7244766	2
4	0.373399472	0.7391245	3
5	0.376632894	0.7450054	7
6	0.377909540	0.7476797	13
7	0.378447688	0.7489294	32
8	0.378698393	0.7495269	70
9	0.378827984	0.7498262	179
10	0.378901312	0.7499830	435
11	0.378945913	0.7500675	1142
12	0.378974611	0.7501136	2947

calculate $H_M(\mathcal{V})$ in an iterative way using the transfer matrix T as

$$H_{M+1}(\mathcal{V}) = \sum_{\mathcal{V}'} T(\mathcal{V}, \mathcal{V}') H_M(\mathcal{V}') \qquad (3.10)$$

The element $T(\mathcal{V}, \mathcal{V}')$ gives the sum (weighted by appropriate powers of z) of all possible ways of going from configuration $\mathcal{V}'$ in column M to configuration $\mathcal{V}$ in column $M + 1$. In figure 3.3 we show a few of the contributions to one particular matrix element (one has to take into account that a given configuration $\mathcal{V}$ stands for a group of actual configurations on the lattice related by symmetry).

Iterating the relation (3.10) one immediately sees that for large R we get

$$G_{0,W}(k, l) \sim \lambda_W(z)^R \qquad (3.11)$$

where $\lambda_W(z)$ is the largest eigenvalue of the transfer matrix. Readers who are familiar with the calculation of correlation functions for spin systems using the transfer matrix may remember that in that case a correlation function is given in terms of the ratio of a lower eigenvalue to the largest eigenvalue. This seems somewhat different from the result (3.11) where only one eigenvalue determines the correlation function. To understand this difference, re-

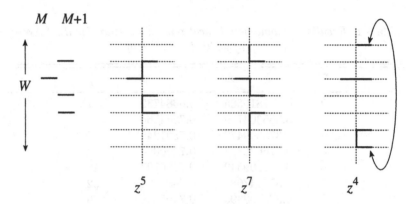

Fig. 3.3. Some contributions to the transfer matrix. On the left, representative states in two neighbouring columns are shown. On the right we show three possible configurations (among others) which can give rise to this situation. Below, we indicate the contribution to the transfer matrix for each of these. Periodic boundary conditions are used.

member that the largest eigenvalue also determines the free energy of a strip for a spin model. From (2.22) it is easy to see that the free energy $f(n)$ of the $O(n)$-model becomes zero when we send $n \to 0$. That implies that the largest eigenvalue is one. And thus we can see that indeed (3.11) is just of the usual form: a ratio of two eigenvalues to the power R.

Comparison of (3.11) with (3.9) then yields the correlation length

$$\xi_W(z) = -\frac{1}{\log \lambda_W(z)} \qquad (3.12)$$

There are two ways to estimate v_c from the correlation lengths. Either we use the data for one particular width and estimate z_c as that value where $\lambda_W(z)$ becomes one. A more interesting way is to use the *phenomenological renormalisation group*. Here one uses the result that under a rescaling by a factor b the correlation length gets a factor b smaller while z is renormalised into z'. For strips of two different widths W and W'

$$\frac{1}{W}\xi_W(z) = \frac{1}{W'}\xi_{W'}(z') \qquad (3.13)$$

The finite size estimate for z_c is then found by looking for a fixed point $z = z'$ of the equation (3.13). The results of such a calculation with $W' = W + 1$ are given in the second column of table 3.1. The extrapolation of these values leads (using the data for odd W) to the estimate $z_c = 0.379052 \pm 0.000007$, to be compared with the result from exact enumerations for $\mu = 1/z_c$ in table 2.3.

As usual in RG calculations the exponent ν can be found from linearising the map $z \rightarrow z'$ around the fixed point

$$\frac{1}{\nu} = \frac{\log \left[(d\xi_W/dz|_{z_c})/(d\xi_{W'}/dz|_{z_c}) \right]}{\log W/W'} - 1 \qquad (3.14)$$

Results for ν from the transfer matrix calculation are given in the third column of table 3.1 and can be extrapolated to $\nu = 0.75018 \pm 0.00008$. This gives strong numerical support for the value $\nu = 3/4$.

Can one also use the transfer matrix to calculate the exponent γ? The answer is yes, but to explain the most powerful way to do that we have to introduce an extra bit of theory.

3.3 Conformal invariance; a fast course

At criticality a physical system is scale invariant. This invariance forms the basis of the renormalisation group, where one rescales the model with a factor b and finds the critical point (loosely speaking) as that point which is fixed (in parameter space) under the transformation. But think how one calculates for example a real space RG transformation. In such a calculation one only uses 'local' information on the spins in a small neighbourhood (this is at least the case when one has a system with short range interactions, if the interactions are long range what we say in this section is not valid). Thus, one can imagine performing a RG calculation where a local, site dependent rescaling $b(\vec{r})$ is used.

What are the consequences of demanding local scale invariance? In $d \geq 3$ few new results are obtained, mainly because in these 'high' dimensions the group of local scale transformations is of finite dimensions. But in $d = 2$, as is known from complex analysis, any conformal transformation is equivalent to a local rescaling. If $w(z)$ is such a conformal mapping, then $|dw/dz|^{-1}$ gives the local rescaling. As this group of transformations is now of infinite dimensions, the requirement of conformal invariance at critical

points is very strong and leads to many new and very unexpected
consequences. These were explored in the 1980s following a seminal paper by Belavin, Polyakov and Zamolodchikov [43]. In the
context of this book we have to limit ourselves to a discussion
of those results which can be derived easily and mention for the
rest only results which are relevant for the subject of this book. A
more extended introduction can be found in reference [44].

~

What do we in fact mean when we say that a model has to be
conformally invariant? This is most clearly expressed at the level
of the correlation functions. Again, let $G(\vec{r}_1, \vec{r}_2)$ be any correlation
function (which at criticality decays with an exponent x) of the
model. Global scale invariance is then expressed mathematically
as

$$G(z_1, z_2) = b^{-2x} G(z_1/b, z_2/b) \qquad (3.15)$$

(where we have now limited ourselves to the case $d = 2$ and have
indicated points in the plane by complex variables z).

In a conformal transformation b depends on z and as a consequence we modify (3.15) into

$$G(z_1, z_2) = |dw/dz(z_1)|^x |dw/dz(z_2)|^x G(w_1, w_2) \qquad (3.16)$$

Demanding that a correlation function (at criticality) transforms
in this particular way leads to many interesting results. In particular, think about the conformal map $w(z) = \frac{W}{2\pi} \log z$ [45]. It maps
the complex plane into a strip of width W. Stated otherwise it
connects the correlation function in the infinite plane to that in a
strip, i.e. to the one which we calculate when performing a transfer
matrix calculation. If we write $w = u + iv$ so that u measures the
distance along the strip, an elementary calculation gives us [45]

$$G(w_1, w_2) = \left[\frac{(2\pi/W)^2}{\left(2\cosh\left(\frac{2\pi}{W}(u_1 - u_2)\right) - 2\cos\left(\frac{2\pi}{W}(v_1 - v_2)\right)\right)} \right]^x$$

In a transfer matrix calculation we are interested in the regime
where $|u_1 - u_2|$ becomes very large and we find in that limit

$$G((u_1, v_1), (u_2, v_2)) \sim \left(\frac{2\pi}{W}\right)^{2x} \exp\left[-\left(\frac{2\pi x}{W}\right)(u_1 - u_2) \right]$$

Comparing with (3.9) then gives the following expression for the correlation length ξ_W^c at criticality

$$\xi_W^c = \frac{W}{2\pi x} \qquad (3.17)$$

From the elementary theory of the transfer matrix we learn that any correlation function can be calculated from a gap Δ in the spectrum of the transfer matrix. The correlation length is just the inverse of this gap. Accordingly, (3.17) becomes

$$\Delta_W^c = \frac{2\pi x}{W} \qquad (3.18)$$

So by calculating the gap in strips of different width we can find the exponent x which describes the decay of the critical correlation function. By studying different gaps we can determine several critical exponents.

Turning back to polymers, we have to remember that in that case a gap corresponds to the difference between the eigenvalue 1 and a particular eigenvalue of the transfer matrix. By calculating, for example, the smallest gap in the spectrum as a function of W we can arrive at the magnetic exponent x_1 and thus at γ. Results of such a calculation are given in table 3.2. Calculations like these for SAWs were first performed in [47].

The extrapolation of these data gives $x_1 = 0.104163 \pm 0.000008$. This has to be compared with the prediction from the Coulomb gas $x_1 = 5/48 = 0.1041666\ldots$. We see that again the transfer matrix very nicely confirms the predictions of the Coulomb gas. Of course we can also determine some of the other watermelon exponents by looking at other gaps in the spectrum of the transfer matrix. Again, good agreement with predictions is found [47].

~

Before concluding this section we inform the reader of some further results coming from conformal invariance, without however being able to derive these results here.

The most remarkable consequence of conformal invariance is that a universality class can be specified by a number called the 'central charge' c [44]. Its precise definition is given in terms of the algebra of infinitesimal generators of the conformal group. For every c the set of possible critical exponents is restricted to a small set of values [48]. In fact, there is close relation with the

Table 3.2. *Results of transfer matrix calculation of the exponent x_1 for the SAW in a strip of width W, from [46]*

W	x_1
1	0.154393
2	0.129184
3	0.116507
4	0.110758
5	0.108099
6	0.106759
7	0.106010
8	0.105551
9	0.105205
10	0.105034
11	0.104879
12	0.104762
13	0.104672

Coulomb gas approach. It is possible to relate c to g. As we know, for every g the set of possible critical exponents is given by (3.3). The set of relevant exponents which we found in the Coulomb gas theory is then found to coincide with those coming from conformal invariance.

How can one determine the central charge of a given model? Again, it can be obtained from an analysis of the spectrum of the transfer matrix [49, 50]. One uses a relation similar to (3.17) which exists for the free energy per unit width f_W of a strip. This quantity is for large W given by

$$f_W = f_\infty - \frac{6\pi c}{W} + \ldots \qquad (3.19)$$

In summary then, here is how in the post-conformal invariance days one goes about studying the critical behaviour of a model in $d = 2$. One begins by setting up a transfer matrix for the model on a strip of width W. Then one finds the critical point of the model using phenomenological renormalisation. In other cases the location of the critical point may be known, as is the case for models showing a self duality, for instance (see chapter 6). At that critical point one determines the central charge of the model from the largest eigenvalue of the transfer matrix using the relation (3.19). Once this is known, one has a list of possible values for critical exponents. At the same time, one calculates the gaps in the spectrum of the transfer matrix, and from these, using (3.18), one gets the critical exponents of the model. Comparing numerical estimates with the set of allowed exponents, it is often possible to conjecture exact values for critical exponents.

Let us now return to the $\widetilde{O}(n)$-model and polymers. It is easy to determine the central charge for polymers. Earlier in this section we mentioned that the free energy of the $O(n)$-model becomes zero when $n \to 0$. This results holds in any geometry, and in particular on a strip. As a consequence, we get $c = 0$ from (3.19). For general n, the central charge $c(n)$ of the $O(n)$-model can be given in terms of n, or using (3.4) in terms of the Coulomb gas parameter g_R [51]. The equation is

$$c(g_R) = 1 - \frac{6(1 - g_R)^2}{g_R} \tag{3.20}$$

Besides critical exponents, conformal invariance can also give results about universal critical amplitudes. This leads to interesting relations also in the case of polymers. We refer interested readers to the literature [52, 53].

3.4 Exact results on the honeycomb lattice

We now discuss how the $O(n)$-model can be solved exactly on the hexagonal lattice [38]. The techniques which are used in this solution have become very popular in the last decade and have led to developments both in statistical physics and in mathematics.

The first step in solving the $O(n)$-model involves a mapping of the model into a so called vertex model. To do this we orient each loop in a graph of the partition function (2.21) of the $O(n)$-model. Each loop can be oriented in two ways, so for a given graph there are 2^L graphs with oriented loops.

We then look at the vertices of the hexagonal lattice, which can be divided into two types (see figure 3.4). At each vertex the oriented loop configuration can be in seven possible states. These

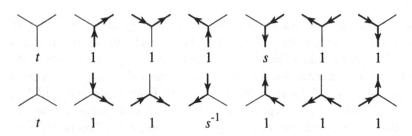

Fig. 3.4. Vertex configurations of the hexagonal lattice $O(n)$-model.
The weight of each configuration is indicated below it (see text).

seven states (each having two possible orientations) are shown in
figure 3.4. It is now our purpose to rewrite the partition function
of the $O(n)$-model in terms of weights which we give to the set of
allowed vertices. To achieve this goal, notice that since there are
only loops in the model, the number of vertices which are visited
by any loop is equal to the number N of edges visited. If we next
introduce the variable $t = 1/v$, and multiply (2.21) by a trivial
factor t^{N_s} where N_s is the number of vertices of the lattice, we
can rewrite the partition function as

$$Z_{O(n)} = \sum_{\mathcal{G}} t^{(N_s - N)} n^L$$

where N is now interpreted as the number of vertices visited by
the loops. It is clear that each empty vertex contributes a factor t
(see figure 3.4). Next, notice that in any oriented loop the number
of turns to the right minus the number of turns to the left is ± 6.
As a consequence if we give each turn to the right a weight w^{-1}
and every turn to the left the weight w, then any oriented loop
gets a weight $w^{\pm 6}$. Since for each loop in the $O(n)$-model there
two differently oriented loops in the vertex representation, we can
rewrite the $O(n)$ partition function in a vertex representation if
we take $n = w^6 + w^{-6}$. However, since all that matters is the total
product of vertex weights it is even possible (figure 3.4) to assign
a weight $s = w^6$ to only one of the vertex configurations in which
the arrow turns to the left. Similarly we assign a weight s^{-1} to
one of the right-turning vertex configurations. In this way each
left-turning loop get a weight s, and each right-turning loop gets
a weight s^{-1}. If we choose s in such a way that $s + s^{-1} = n$ we

have rewritten the $O(n)$ partition function in terms of all possible configurations of vertices on the hexagonal lattice. Each configuration gets a total weight which is the product of the weights of all individual vertices.

In terms of t, the criticality condition (3.7) is rewritten as

$$n = 2 - (2 - t^2)^2 \qquad (3.21)$$

As mentioned, (3.7) holds for $-2 \leq n \leq 2$. It is therefore common to parametrise the critical line with a parameter α as follows

$$t^2 - 2 = 2\sin 3\alpha \qquad (3.22)$$

The criticality condition (3.21) then becomes $n = 2\cos 6\alpha$. For s we get $s = \exp(i6\alpha)$.

The advantage of this mapping is that a lot is known about how to solve vertex models using a transfer matrix techniques [55]. This approach started with the work of Lieb [54] on the square lattice six vertex model.

Baxter [38] was able to figure out how the vertex representation of the $O(n)$-model can be solved with the transfer matrix. We cannot discuss this in all detail, but we will sketch the main idea. The first step consists in redrawing edges in terms of different types of lines (dotted, full, wiggly, etc.) as indicated in figure 3.5. An up pointing arrow is replaced by a dotted line, etc. As a consequence the vertex configurations are transformed into the configurations of figure 3.6. The main observation to be made is that the total number m of full and wiggly lines is conserved when, on the hexagonal lattice, one goes from one row of vertical edges to the next one. Let us now set up a vertical row to vertical row transfer matrix for the hexagonal lattice. We denote by $\mathcal{V}$ the configuration of a particular row and by $\mathcal{V}'$ the configuration of the next row of vertical bonds. As was the case for the SAW the transfer matrix $T(\mathcal{V}, \mathcal{V}')$ is given by a sum of all possible configurations of horizontal rows that allow one to go from $\mathcal{V}$ to $\mathcal{V}'$. Each term in this sum is weighted by the product of all vertex weights in that horizontal row. Now, because m is conserved the transfer matrix consists of blocks in which m is fixed. The next step is to find the eigenvalues for each block. For $m = 1$ this is not too difficult. Then one uses a Bethe *Ansatz* to guess eigenfunctions for general m as particular combinations of those for $m = 1$ (in a way analogous to what we did in section 1.4). It turns out that this trial

Fig. 3.5. Conversion of arrow configurations (after reference [37]).

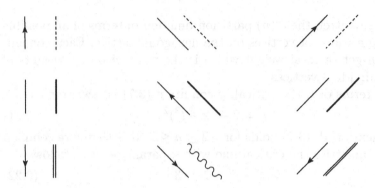

Fig. 3.6. Vertex configurations of the $O(n)$-model after the conversions of figure 3.5 have been made (adapted from reference [37]).

leads to the correct solution if one is along the critical line (3.21) of the $O(n)$-model. In his work, Baxter determined the largest eigenvalue of the transfer matrix as a function of n, which is sufficient to obtain the free energy of the model in the thermodynamic limit. In a subsequent study, Batchelor and Blöte [39] extended these calculations to strips of finite width W from which, using the results of conformal invariance (in particular equations (3.18) and (3.19)), one can determine the scaling exponents and the central charge. These calculations have fully confirmed the values of the watermelon exponents and the central charge as determined from the Coulomb gas and conformal invariance methods.

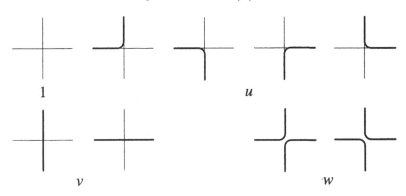

Fig. 3.7. Vertex configurations for the square lattice $O(n)$-model.

3.5 A square lattice $O(n)$-model

After the exact solution on the honeycomb lattice, a new $O(n)$-like model was introduced by Nienhuis on the square lattice [56]. Such a model can be of interest in investigating, for example, universality questions. More interestingly, the square lattice model has a very rich structure and can be related to such problems as dense polymers (chapter 7) and polymers at the θ-point (chapter 8).

The square lattice $O(n)$-model can be defined in analogy with the honeycomb model (2.17). In this case however the spins are located on the edges of the lattice. We will denote by $\vec{s}_{1,i}, \ldots, \vec{s}_{4,i}$ the four n-dimensional spin vectors located on edges incident on vertex i (counted anticlockwise). The partition function $Z_{\square,n}(u, v, w)$ of the model is defined as

$$Z_{\square,n}(u, v, w) = \text{Tr} \prod_i Q_i \qquad (3.23)$$

where Q_i is given by

$$
\begin{aligned}
Q_i = \quad & 1 \quad +u[\vec{s}_{1,i} \cdot \vec{s}_{2,i} + \vec{s}_{2,i} \cdot \vec{s}_{3,i} + \vec{s}_{3,i} \cdot \vec{s}_{4,i} + \vec{s}_{4,i} \cdot \vec{s}_{1,i}] \\
+ \quad & v[\vec{s}_{1,i} \cdot \vec{s}_{3,i} + \vec{s}_{2,i} \cdot \vec{s}_{4,i}] + w[(\vec{s}_{1,i} \cdot \vec{s}_{2,i})(\vec{s}_{3,i} \cdot \vec{s}_{4,i}) \\
+ \quad & (\vec{s}_{2,i} \cdot \vec{s}_{3,i})(\vec{s}_{1,i} \cdot \vec{s}_{4,i})] \qquad (3.24)
\end{aligned}
$$

The various terms in this sum are represented schematically in figure 3.7. One can expand the product in (3.23) and after performing the trace obtain a loop gas on the square lattice whose partition function is given by

$$Z_{\Box,n}(u,v,w) = \sum_{\mathcal{G}} u^{N_u} v^{N_v} w^{N_w} n^L \qquad (3.25)$$

where the allowed graphs contain L closed loops on the lattice which contain respectively N_u, N_v and N_w vertices of weight u, v and w. Notice that now the loops occurring in the expansion of the partition function are not necessarily self avoiding.

Batchelor, Nienhuis and Warnaar [57, 58] have been able to solve this model in a way similar to that described in the previous section for the $O(n)$-model on the hexagonal lattice; firstly the model is translated into a vertex model (which, as can be verified by the reader, in this case is a 19-vertex model). Secondly one looks for lines along which the corresponding vertex model can be solved with the Bethe *Ansatz* technique. For the model (3.25) Blöte and Nienhuis [59] determined four branches in the parameter space where such a solution is possible. These branches are parametrised as follows

$$\begin{aligned} u &= 4w \sin(\theta/2) \cos(\pi/4 - \theta/4) \\ v &= w[1 + 2\sin(\theta/2)] \qquad\qquad (3.26) \\ w &= \frac{1}{2 - [1 - 2\sin(\theta/2)][1 + 2\sin(\theta/2)]^2} \end{aligned}$$

θ is related to n by

$$n = -2\cos(2\theta) \qquad (3.27)$$

It is clear that θ plays a role similar to that of g_R in the Coulomb gas. Since we focus here on the exact critical behaviour of the model the relation with the Coulomb gas is irrelevant. The four branches along which the model can be solved exactly correspond to the cases

$$\begin{array}{lll} \pi/2 \leq \ \theta \ \leq \pi & \text{branch 1} \\ 0 \leq \ \theta \ \leq \pi/2 & \text{branch 2} \\ -\pi/2 \leq \ \theta \ \leq 0 & \text{branch 3} \\ -\pi \leq \ \theta \ \leq -\pi/2 & \text{branch 4} \end{array}$$

As is the case for the honeycomb $O(n)$-model, the Bethe *Ansatz* then allows the exact calculation of the free energy per vertex. In a series of papers, Batchelor, Nienhuis and Warnaar [57, 58] were able, using many sophisticated techniques, to solve the model on strips of finite width and thus to determine also the central charge and the spectrum (and thus the critical exponents) of this model along the four branches. For us it is enough to note that along branch 1 the central charge and critical exponents of the hexagonal lattice model are recovered, thus showing that along this branch the square lattice $O(n)$-model is in the same universality class as the more traditional model. The critical behaviour along one of the other branches can be identified as that of the low temperature phase of the $O(n)$-model (chapter 7). For a complete discussion of the critical behaviour of the model the reader should consult the original references.

Finally, it is of interest to remark that there exists a fifth branch (actually called branch 0) along which the model is solvable. This is the branch given by $u = w = 1/2, v = 0$, i.e. along this branch the location of the critical point doesn't depend on n. We will see that this branch shows up when studying SAWs on the Manhattan lattice (chapters 7 and 8). The square lattice model along this branch was solved exactly by Batchelor [60] by first mapping the model onto a 15 vertex model (since $v = 0$) and then solving this vertex model with Bethe *Ansatz* techniques.

3.6 The SAW on fractal lattices

In the previous sections we have seen how to obtain many exact results about the SAW in $d = 2$. We conclude this chapter by discussing how the SAW can also be solved exactly on some fractal lattices. This is one of the rare situations in which the renormalisation group (RG) can be worked out exactly.

The fractal which we will study in particular is the Sierpinski gasket, the construction of which is shown in figure 3.8. At the n-th level of construction, the number N_n of sites in the gasket is

$$N_n = \frac{3}{2}(1 + 3^n)$$

Since at each step of the construction the elementary length is reduced by a factor $b = 2$, the fractal dimension D of this gasket is $D = \log 3 / \log 2$.

The study of the SAW on deterministic fractals was initiated in the work of Dhar [61]. The basic idea is a recursive construction of all possible SAWs between two points on the gasket. Consider

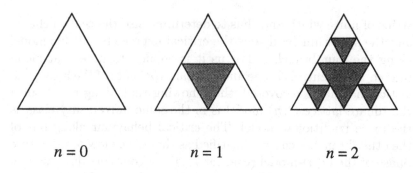

$n = 0$ $n = 1$ $n = 2$

Fig. 3.8. Construction of the Sierpinski gasket with $b = 2$.

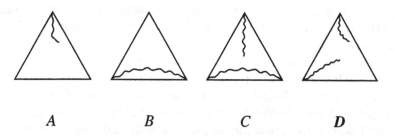

A   *B*   *C*   *D*

Fig. 3.9. Restricted partition functions for the SAW on the Sierpinski gasket.

a triangle of the Sierpinski gasket at order r. In figure 3.9 we show the possible configurations of a SAW in such a triangle. For example, the situation in A shows the case where the SAW starts in the particular triangle and goes out along one of the corners. In situation C, the SAW starts in the triangle, goes out, but comes in a later stage through another corner to go out again at the third corner. At level 0 the situations depicted in C and D cannot occur. Following Dhar, we will associate a fugacity v with each *vertex* visited by the walk. The vertices at the first and last step are given a weight $\sqrt{v}$. In this way, each SAW gets its correct weight. Therefore, at level 0 the weight of each of the configurations is

$$
\begin{aligned}
A_0 &= \sqrt{v} \\
B_0 &= v \\
C_0 &= 0 \\
D_0 &= 0
\end{aligned}
$$

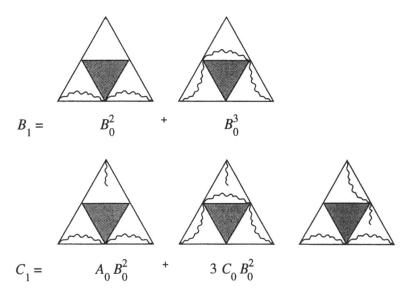

$$B_1 = B_0^2 \quad + \quad B_0^3$$

$$C_1 = A_0 B_0^2 \quad + \quad 3 C_0 B_0^2$$

Fig. 3.10. Graphs contributing to the renormalisation of B and C.

Let us go one level further. If a triangle is in configuration B at level 1, it can be in two possible configurations at the lower level (figure 3.10). One therefore obtains

$$B_1 = B_0^2 + B_0^3$$

For A_1 we get in a similar way,

$$A_1 = A_0(1 + 2B_0 + 2B_0^2)$$

The general recursion relation for going from level r to $r + 1$ can also be written down easily. We quote the results for B, A and C and leave the other as an exercise for the reader

$$
\begin{aligned}
B_{r+1} &= B_r^2 + B_r^3 \\
A_{r+1} &= A_r(1 + 2B_r + 2B_r^2) + C_r(2B_r^2) \\
C_{r+1} &= A_r B_r^2 + 3C_r B_r^2
\end{aligned}
\tag{3.28}
$$

Notice that the iteration of B does not involve any of the other variables. From these recursions it is possible to obtain the critical behaviour of the SAW using RG techniques. Indeed we can interpret the recursion for B as an RG equation for v. At any level r, B_r just gives the weight of all walks going between two corners of an r-th level triangle. This quantity is interpreted as the fugacity for a step at that level. In this way, we can rewrite the recursion for B as an RG equation for v

$$v' = v^2 + v^3 \qquad (3.29)$$

The flow of this RG equation is very simple. There are two attractive fixed points (at $v = 0, v = \infty$) and one repulsive fixed point at

$$v_c = \frac{1}{2}(\sqrt{5} - 1) \qquad (3.30)$$

From this we obtain immediately the connective constant on the Sierpinski gasket

$$\mu_{SG} = \frac{2}{\sqrt{5} - 1} \approx 1.6180 \qquad (3.31)$$

To determine ν we take the derivative of v' at v_c

$$2^{1/\nu} = 2v_c + 3v_c^2$$

from which we get $\nu \approx 0.7986....$ To obtain γ we need to use the recursion for A and, since they are coupled, also that of C. These recursions are like RG equations for symmetry breaking fields in spin models. They have a fixed point at $A = C = 0$. We therefore take the matrix of derivatives at the fixed point $B = v_c, A = C = 0$ and calculate its eigenvalues. The largest eigenvalue λ_H is then

$$\lambda_H = \frac{[3 + 3v_c^2 + (9 - 18v_c^2 + 17v_c^4)^{1/2}]}{2}$$

From λ_H we get the magnetic exponent y_H as $2^{y_H} = \lambda_H$. Finally, assuming hyperscaling we can obtain γ

$$\gamma = \nu\left(-D + 2\frac{\log \lambda_H}{\log 2}\right) \approx 1.3752 \qquad (3.32)$$

While one can argue that the SAW on a Sierpinski is a somewhat academic problem, it has many useful properties. Firstly, it is a case in which a RG can be performed exactly, without the need of uncontrolled approximations as is so often the case when using the RG (in particular in real space). Secondly, the study of a particular model on the Sierpinski gasket or other deterministic fractals may give a qualitative insight into the behaviour of that model on a Euclidean lattice, where often no exact calculation can be performed. Finally, it may be useful as a first step in the study of a model like the SAW on a random fractal.

Following the work of Dhar the SAW has been studied on several other fractals. Particularly interesting work in this respect was performed by Elezović *et al.* [62] who studied the SAW on a whole class of Sierpinski-like fractals, which are parametrised by the rescaling factor b. For $b \to \infty$ the fractal dimension of these gaskets goes to 2, so the original idea was that exponents like γ would approach their value for $d = 2$ when $b \to \infty$. To their surprise, Elezović *et al.* found that this seems not to be the case (they went up to $b = 8$). In later work, Dhar [63] showed, using the exact results from Nienhuis for the SAW on the two-dimensional lattice and finite size scaling theory, that $\gamma(b \to \infty) = 133/32$ instead of the two-dimensional value $\gamma = 43/32$.

In recent years, several of the problems which we will discuss in further chapters (such as polymer adsorption and polymer collapse) have been studied on the Sierpinski gasket.

$$\sim$$

To summarise this chapter, we have seen how through the relation with the $O(n)$-model, the use of the Coulomb gas, conformal invariance, exactly solvable models, and numerical verification using series analysis and especially transfer matrix calculations, one has nowadays a very detailed knowledge of the critical properties of the SAW in $d = 2$. The same can be said about the behaviour of the SAW on deterministic fractal structures.

Since we know that in $d \geq 4$ SAWs have mean field critical behaviour, we are left with one more dimension to study, which (maybe unfortunately) is also the most relevant one for real polymers. So, let's turn our attention to three dimensions.

4

The SAW in $d = 3$

In $d = 3$ neither conformal invariance nor the Coulomb gas technique is very helpful in determining the critical behaviour of self avoiding walks. The transfer matrix can only reach up to small widths W, series can only get up to rather small N, and so on. So we have to look for different methods. There are essentially three of these. The first is probably the most obvious one; we can perform experiments on real polymers. In this book, we will only mention the results of these. Secondly, a powerful numerical method which so far has not been discussed is the Monte Carlo technique. It can of course be applied more easily and more accurately in $d = 2$, but in $d = 3$ it has less competition from other methods. That's why we will discuss it in section 4.2 mainly from the point of view of learning about polymers in $d = 3$. We begin the chapter with a brief discussion of the third method, which is the RG approach to the critical behaviour of polymers.

4.1 Direct renormalisation of the Edwards model

We already encountered the exact RG methods for SAWs on fractal lattices in the previous chapter. But such a real space approach can only work very approximately on Euclidean lattices [64]. The most precise RG calculations for polymers are performed with continuum techniques. A first method uses the $O(n)$-model, calculates the exponents of that model using techniques such as the ϵ-expansion, and then sends $n \to 0$ in the final equations. Here we will not get into these calculations; they have been very well described elsewhere [65].

Instead we will give a very brief overview of the 'direct renormalisation' method [66]. The idea is to start from a continuum expression for the partition function of the SAW, something like

(1.12) for random walks (which can of course be evaluated exactly in that case) and then to calculate the critical behaviour using the RG.

A continuum model that was introduced to describe the same physics as that of the SAW is the *Edwards model* [67]. It is defined using a path integral which gives the weight for each possible path between two points in 'space-time'

$$P(\vec{x}, t; \vec{0}, 0) = \int \mathcal{D}\vec{y}(t) \exp \left[- \frac{1}{4D} \int_0^t \dot{\vec{y}}^2(t') dt' \right.$$
$$\left. - \frac{b}{2} \int_0^t \int_0^t \delta(\vec{y}(t') - \vec{y}(t'')) \, dt' dt'' \right] \quad (4.1)$$

When $b = 0$ we recover the Brownian paths encountered in chapter 1 (we have set $\vec{x_0} = t_0 = 0$). Each of the paths gets a weight in which self encounters are punished. We can expect that for $b \to \infty$ only paths which do not contain self intersections survive. That's why we believe that this Edwards model describes the same physics as the SAW.

One similarity is already obvious from a dimensional analysis. Let us rescale time by a factor t $(t' = t\lambda)$ and length by a factor $\sqrt{t}$ $(\vec{y} = \sqrt{t}\vec{x})$. From what we learned in the first chapter such a rescaling leaves the first term in the exponent of (4.1) invariant. Using the well known property of the Dirac-δ, $\delta(ax) = \delta(x)/|a|$, we see that under this rescaling the second term (we will denote this term by B) in the exponent of (4.1) is modified into

$$B = \frac{b}{2} t^{2-d/2} \int_0^1 \int_0^1 \delta(\vec{x}(\lambda) - \vec{x}(\lambda')) d\lambda d\lambda'$$

Notice that the integrals no longer depend on t. From this equation we immediately see that for long walks $(t \to \infty)$ the interaction term becomes unimportant, when $d > 4$. This is just what we already know to be true for SAWs.

How does one continue to evaluate (4.1)? The usual way to do this is to make a Taylor expansion of the second term in the exponential. Each resulting term in this expansion is then a Gaussian average over a power of B. The actual calculation involves ideas like regularisation (to calculate the integrals in B which would otherwise diverge) and renormalisation (to extract exponents from the series for $P(\vec{x})$ or from similar expansions for the average end-to-end distance). The renormalisation itself uses an expansion in

$\epsilon = 4 - d$. The details of these calculations have been described well in [33] and [65]. We just quote here the results of a calculation to first order

$$\gamma = 1 + \epsilon/8 + \dots \tag{4.2}$$
$$\nu = 1/2 + \epsilon/16 + \dots \tag{4.3}$$

For $\epsilon = 0$ we recover the well known RW exponents. The series for ν has been calculated up to order ϵ^5 [33]. One can furthermore apply resummation techniques. All these extensive calculations lead to the final estimate in $d = 3$ [68]

$$\nu = 0.588 \pm 0.0015 \qquad \gamma = 1.157 \pm 0.003 \tag{4.4}$$

This can be compared with the experimental result coming from neutron diffraction studies, $\nu = 0.586 \pm 0.004$ [65].

4.2 Monte Carlo methods

In order to calculate thermodynamic properties of SAWs we have to perform an average over the set of c_N different configurations in which the polymer can be. Since c_N grows exponentially fast, performing this average exactly becomes impossible above a certain value of N. The idea of a Monte Carlo method is to obtain a representative sample of the set of configurations and to calculate averages over this restricted set (a general reference to Monte Carlo methods for polymers is [69]). If the set is well chosen one can hope to get accurate approximations to real averages. Sure enough, the bigger the sample becomes the better the accuracy.

There are by now a large number of methods of generating sets of SAWs. Here we will consider a few of them. The simplest method is to generate a random walk on (say) the hypercubic lattice, and to check at each time step whether the walk is self avoiding. If not, we throw away the walk we have generated up to that step. Otherwise we continue until we reach the desired length. It is obvious that owing to self trapping it will be difficult to generate long walks. In fact only exponentially few of all random walks are self avoiding, so this method is not very efficient. A variant of this method is what we can call a myopic self avoiding walker. It was introduced by Rosenbluth and Rosenbluth [70] and therefore often also goes under the name of the Rosenbluth

Fig. 4.1. Growth of the myopic SAW.

method. It goes as follows. After making an initial step the walker 'looks' around to determine the number of nearest neighbour sites which up to that moment it has not visited (this is the reason the walker is called myopic). It then chooses completely at random one of these sites and moves to that place. In this way the walker continues until it has reached a site from where it can no longer continue or until it has reached the maximum number of steps one wants to investigate. In this way we generate a set of SAWs. One has to realise that in this way not all SAWs of a given length are created with the same probability. Consider the two SAWs of five steps shown in figure 4.1. After the third step, the walk on the left can only choose between two possibilities to continue, whereas the walk to the right can choose between three possibilities at each step. Thus, the walk on the left will be generated with a probability 3/2 times higher than that of the walk on the right. Therefore to calculate averages in an appropriate way we have to associate with each walk i a weight W_i. In the case of the walk to the left this would be 2/3.

Suppose then we generate a_N SAWs of N steps in this way. An estimate of the average over the set of SAWs of a quantity A can be calculated as

$$\langle A \rangle \approx \frac{\sum_{i=1}^{a_N} A_i W_i}{\sum_{i=1}^{a_N} W_i} \qquad (4.5)$$

Unfortunately, in this case too the number of successfully generated SAWs goes down exponentially with N. There is also no obvious way to determine c_N for this method. Finally, it is almost impossible to generate walks of a fixed topology; for instance, it is impossible to study SAPs with this method since so few of them are generated. But this method is very simple to program, and the walks one generates are really independent and (taking into account the weights W_i) are uniformly distributed over the set of all SAWs of given length.

~

A second class of method can be called dynamic; that is, they change a given SAW into another one. Here we do not imply in any way that the dynamics used are a realistic copy of the dynamics of the polymer, although for some methods this can be the case. These methods are based on the theory of Markov processes. They set up a 'random walk' in the space of SAW configurations and the idea is to generate such a chain that eventually (after many steps of the algorithm) one visits each element of the state space with equal probability. If that is achieved, averages over the space of states can be calculated as time averages over the history of the walk (ergodicity). The SAWs which are generated in consecutive steps of the Markov chain are however correlated, so that care must be taken in calculating these time averages. To fix attention, let us discuss the 'pivot' method, introduced by Lal [71] in 1969. Originally, the method attracted little attention, but after a detailed investigation by Madras and Sokal [72] it has gained a lot in popularity. In this method one starts from a given SAW and then picks at random one site along the walk; this is called the pivot site. It divides the walk into two parts. Then we perform one of the lattice symmetry operations (like reflection or rotation) to the second part of the SAW (more details are given below). The new walk generated by this procedure can be self avoiding, in which case we accept it, or not. In the latter case the newly formed walk is rejected, and the 'new' walk is equal to the original walk. This procedure is then iterated.

In d dimensions the group of transformations that leave the hypercubic lattice invariant has $2^d d!$ elements. The factorial term gives the number of ways in which we can permute the different coordinate axes. For a given permutation of the axes, each unit vector can point in two directions. This gives the factor 2^d. Figure 4.2 shows the eight possible moves that are allowed in $d = 2$. What is the rate of success; i.e. what fraction of newly generated SAWs gets accepted? A rough estimate can be obtained as follows [72].

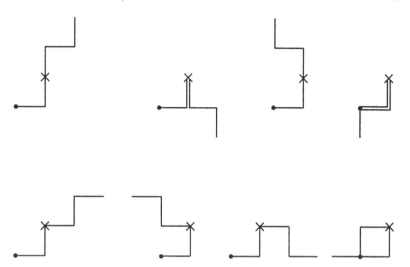

Fig. 4.2. The two-dimensional pivot algorithm. Starting from the SAW on the top left, we can generate seven other SAWs by performing all moves allowed by the algorithm. Three of these lead to walks that are not self avoiding. The dot indicates the starting point of the SAW, the cross indicates the location of the pivot point.

Suppose we start from an N-step SAW and that the pivot is at step m. In this way our original SAW is divided into two SAWs, one of length $N - m$, the other of length m. After performing the transformation on the second part of the walk, attaching it again to the first part of the walk is somewhat similar to concatenating an m-step and an $(N - m)$-step SAW. The fraction of successes $f_{\rm s}(N)$ is therefore approximately given by

$$f_{\rm s}(N) \approx \sum_{m=1}^{N} \frac{c_N}{c_m \, c_{N-m}} p_m$$

where p_m is the probability that the pivot is found at m. Filling in the by now well known expression for c_N we get the estimate

$$f_{\rm s}(N) \approx N^{-\gamma+1} \sum_{m=1}^{N} p_m \left(\frac{N}{m}\right)^{\gamma-1} \left(\frac{1}{1 - m/N}\right)^{\gamma-1} \qquad (4.6)$$

The sum has at most a very weak dependence on N so we see that $f_{\rm s}(N) \sim N^{-p}$ whence we estimate $p = \gamma - 1$. Notice that $f_{\rm s}(N)$

decreases much more slowly for the pivot algorithm than for the algorithms we discussed earlier where the decrease is exponentially fast. Of course, this estimate for p is rather crude. Detailed numerical investigations with the pivot algorithm usually give a somewhat larger value of p, but are still consistent with a power law decay for $f_s(N)$.

In summary, running the pivot algorithm for a long time we get a subset of the set of all N-step SAWs. If we knew that the pivot algorithm was ergodic, visiting all SAWs with equal probability, we could obtain estimates for average properties of SAWs by just averaging over this subset. So, let's turn to the question of ergodicity. Here we have to use some general results from the theory of Markov chains on finite-dimensional spaces [73]. Let W and W' represent two SAWs and denote by $P(W, W')$ the probability that a step in a Markov chain algorithm, e.g. one step of the pivot method, transforms W into W'. Furthermore, let $P^n(W, W')$ be the probability that W is changed into W' after n steps of the algorithm. Suppose that for each W and W' there exists an $n \geq 0$ such that $P^n(W, W') > 0$. In that case the Markov chain is said to be *irreducible*. Secondly let $\pi(W)$ be a left eigenvector of P with eigenvalue 1, i.e. $\pi(W)$ satisfies

$$(a) \quad \pi(W) \geq 0 \;\; \forall W, \qquad \sum_W \pi(W) = 1 \qquad (4.7)$$

$$(b) \quad \sum_W \pi(W) P(W, W') = \pi(W') \qquad (4.8)$$

Then it can be shown that for $n \to \infty$ the probability that the Markov chain produces walk W is just given by $\pi(W)$. Time averages thus becomes averages over the distribution π. Thus if we choose our transition probabilities $P(W, W')$ in the right way (so that they satisfy (4.8)) we can generate a chain of walks with any desired probability distribution. A way to satisfy (4.8) is known as the *detailed balance* condition. This condition is

$$\pi(W) P(W, W') = \pi(W') P(W', W) \qquad (4.9)$$

Summing (4.9) over W gives (4.8). Thus, if we choose transition rates that satisfy detailed balance our Markov chain will 'visit' each SAW with a probability distribution π that is determined by (4.9). The well known Metropolis algorithm for spin systems is based on such a detailed balance condition using the Boltzmann weights. In our case, we want to generate SAWs with a uniform distribution, i.e. for N-step SAWs we want $\pi(W) = 1/c_N$. Detailed balance teaches us that this can be achieved by choosing transition rates that are reversible, i.e. satisfy $P(W, W') = P(W', W)$. Madras and Sokal [72] derived the conditions for the pivot algorithm to be ergodic. It turns out that one can leave out some of the $2^d d!$ transformations and still keep ergodicity.

Finally, we turn briefly to autocorrelations introduced in the Markov chain. SAWs that are generated in consecutive steps of a Markov chain are in general correlated. But after some time τ_a, called the autocorrelation time, these correlations have decayed sufficiently that if we take SAWs generated at multiples of τ_a, we can estimate averages and errors using standard statistical techniques for a set of independent data. It is difficult to get any hard results on τ_a. Since a step in the pivot algorithm, when accepted, makes in general a big change in a SAW, it is believed that for global properties τ_a is a couple of times $1/f_s(N)$. This estimate should work well for properties such as the end-to-end distance. If one is interested in local properties, τ_a may become much larger (see [72] for a detailed discussion).

At this moment, the pivot algorithm is probably the most precise and best understood method of investigating very long SAWs. It has been used recently by Li, Madras and Sokal [74] in a very extensive study of the critical properties of SAWs in $d = 2$ and $d = 3$. In $d = 2$ these calculations again confirm the value $\nu = 3/4$. In $d = 3$ the SAWs studied have lengths up to 80,000. Such extensive calculations lead to the estimate

$$\nu = 0.5877 \pm 0.0006 \tag{4.10}$$

which compares extremely well with the results quoted in (4.4). Earlier Monte Carlo data gave a somewhat higher value for ν. It is now clear that this was a result of rather strong corrections to scaling. Indeed, so far we have discussed in this book only the leading terms in the behaviour of $\langle R_N^2 \rangle$ as given by (1.4). Correction terms are of power law form

$$\langle R_N^2 \rangle = A_0 N^{2\nu}(1 + A_1 N^{-\Delta_1} + \ldots) \tag{4.11}$$

In $d = 3$, the work of Li *et al.* leads to the estimate $\Delta_1 = 0.56 \pm 0.03$ while in $d = 2$ one predicts the exact value $\Delta_1 = 3/2$ (this value was first predicted in [32]; numerical evidence for this value is reported in [19] and [75]).

We will just mention two more properties of the pivot method. Firstly, there is no good way to estimate μ or γ using the pivot algorithm. Indeed, Monte Carlo methods are in general quite good in estimating average values such as $\langle R_N^2 \rangle$ but are less successful when one has to estimate quantities such as partition functions. Secondly, there exists an extension of the pivot algorithm which is useful in studying SAPs [76]. This method starts from a given SAP and then picks two distinct pivot points along the polygon. Then a transformation is applied to the part of the SAP between the two pivots. For details we refer the reader to [76]. One can show that in this case also, the algorithm is ergodic. The study of SAPs is especially relevant when one is interested in topological properties of polymers (see chapter 10).

Besides the simple growth and the pivot algorithm, many more Monte Carlo techniques for polymers exist. We conclude this section by a brief discussion of some other methods which have gained some popularity in recent years, and which we will encounter again later in this book.

The first of these methods is known as the BFACF algorithm (after its inventors, Berg, Foerster, Aragão de Carvalho, Caracciolo and Frölich [77, 78]). Like the pivot, the method uses a Markov chain to generate a set of SAWs. But in contrast to the pivot the method works in a grand canonical ensemble which means that the length of the SAW is not fixed and that we have to introduce a fugacity z which controls its average length. The elementary moves of the algorithm are shown in figure 4.3. In the move on the left hand side of the figure the number of steps in the SAW is changed by ± 2. In the move on the right hand side the length of the polymer isn't changed. Each of the possible moves has an associated probability denoted by p_+, p_- and p_0. In the BFACF algorithm one starts from a given SAW. We pick at random one of the bonds of the walk. There are $2d - 2$ directions perpendicular to that of the chosen bond. For each of these exactly one of the elementary moves can be performed. This gives rise to $2d - 2$ new walks, which don't necessarily have to be self avoiding. For each of the moves, we add up the corresponding probability. If this is less than 1, we add as the $(2d - 1)$-th walk the walk from which

Fig. 4.3. Elementary moves in the BFACF algorithm.

we started. Now we choose one of the walks out of the set of $2d-2$ (or $2d-1$) newly generated walks (taking into account probabilities). If the newly formed walk is again a SAW it will be the next SAW in the Markov chain; if it isn't self avoiding we keep the previous walk. This procedure is then iterated. It is obvious that this local algorithm moves more slowly through the space of SAWs than does the pivot algorithm. Therefore its autocorrelation time is expected to be much larger. Notice that this algorithm never changes the endpoints of the SAW.

What can we say about the properties of this algorithm? First, one can show that when the probabilities p_+, p_- and p_0 obey certain constraints, the transition rates $P(W, W')$ obey detailed balance for the 'grand canonical weight' $\pi(W) \sim N(W)z^{N(W)}$ [79]. Here $N(W)$ is the length of the walk and we have to keep in mind that we can only sample the subspace of SAWs with given fixed endpoints. The above mentioned constraints together with the demand to make the autocorrelation time as small as possible leads to an 'optimal' choice for the transition probabilities which is

$$p_+ = \frac{z^2}{1 + (2d-3)z^2}$$

$$p_- = \frac{1}{1 + (2d-3)z^2}$$

$$p_0 = \frac{1 + z^2}{2[1 + (2d-3)z^2]}$$

In $d = 2$ the algorithm can also be proven to be irreducible within the space of all SAWs with given fixed endpoints. Then, when the transition rates are chosen to obey detailed balance, the algorithm is ergodic. In $d = 3$ the algorithm is not always irreducible, especially not for SAPs. This is mainly due to the possible presence of knots. On the other hand, this is exactly one of the interesting properties of the algorithm we shall discuss in chapter 10.

Another interesting aspect of the BFACF algorithm is that it allows a determination of μ and α. Let the distance between the fixed endpoints of the SAW be r. Then, for the $\pi(W)$ of the BFACF algorithm, the fraction of time that the walk has a length N is given by

$$\frac{N c_N(r) z^N}{\sum_N N c_N(r) z^N}$$

Now we can use our result (2.35) for $c_N(r)$, from which it follows that in the regime $r << N^\nu$, $c_N(r) \sim N^{\alpha-2}$ (or stated physically, on these length scales, the SAW looks like a polygon). Inserting this into the above equation we find that at fixed value z and for a set of SAWs with fixed endpoints the time spent in configurations of length N is proportional to $(\mu z)^N N^{\alpha-1}$. So by measuring the times the SAW has length N we can get μ and α. Another method that works in the grand canonical ensemble and is even simpler than the BFACF method is known as the Berretti–Sokal method [80]. In this method one deletes the last step of the SAW with a probability p or adds a new bond at the end of the SAW with a probability $1 - p$. If this latter possibility is chosen, one picks one of the $2d$ neighbours of the N-th monomer and adds a bond to that neighbour. The step is accepted if the resulting walk remains self avoiding. The probability p is chosen in a way that the probability of generating a walk of length N is proportional to z^N. In particular, one takes

$$p = \frac{1}{1 + 2dz}$$

As in the case of the BFACF method, the Berretti–Sokal algorithm allows the determination of c_N.

We conclude our discussion of Monte Carlo methods for polymers on a lattice by introducing the so called bond fluctuation model [81], introduced by Carmesin and Kremer. In this model each monomer is represented by an elementary hypercube of the d-dimensional lattice. Figure 4.4 illustrates the situation. A step of the algorithm consists of moving one such hypercube to a near-

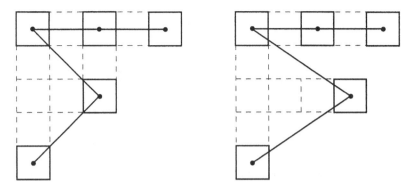

Fig. 4.4. An allowed move in the simulation of the bond fluctuation model.

est neighbour cube, taking care that self avoidance is maintained. In such a move, the length l of the bond between two consecutive monomers along the chain may change; that is the origin of the name of the model. However, we restrict the length to values less than $\sqrt{13}$ (measured in lattice units) in $d = 2$. In $d = 3$ we require $l \leq \sqrt{10}$. If one starts with a configuration that doesn't contain bond crossings, these restrictions on l guarantee that crossing never occurs in the time evolution of the polymer. This model is a first step between a lattice model and a continuum model since both the allowed lengths and the allowed bond angles take on more values than for a SAW. Of course, the fact that l fluctuates is physical, since we must always remember that our description of the polymer is on a coarse grained level in which a 'monomer' stands for a collection of real monomers. It is therefore realistic to assume that the bond lengths and angles fluctuate. In some calculations, a potential controlling the length and bond angles is included. The bond fluctuation method can be programmed more effectively than continuum models since one still works on a lattice, which is suitable in implementing the model. The model also gives a more realistic description of the real dynamics of a polymer. Combined with the fact that it works well for dense polymer systems, this explains why the model has gained a lot of popularity in recent years, even though the method is not ergodic.

5

Polymers near a surface

In the previous three chapters we considered the behaviour of polymers in bulk. In reality systems are never infinite and one always has to consider the presence of surfaces. When the polymer is close to or even attached to a surface its critical properties may change. When there is an interaction between the monomers and the surface interesting *adsorption* effects can occur. We now turn to a discussion of these phenomena.

5.1 Surface magnetism

Consider a $(d > 1)$-dimensional lattice, in which a polymer is restricted to be in a semi-infinite region, e.g. the region with $x \geq 0$. We imagine a wall at $x = 0$ which is impenetrable. When the polymer is very far from this surface, i.e. in the bulk, its properties are hardly changed by the presence of the wall. As the polymer is placed closer and closer to the surface its properties may be modified. These effects can be expected to have a scaling behaviour depending on the ratio of the distance x_{cm} of the centre of mass of the polymer from the surface and its radius R_N. As we are interested in properties of the polymer on a coarse grained scale, let us immediately attach one of the monomers to the surface (figure 5.1). There are two ways to discuss the properties of such a polymer. The first one works directly with SAWs, while the other one uses, through the $O(n)$-connection, known properties of surface magnetism. Since we don't expect the general reader to know about surface critical behaviour, we will give a brief overview of the main ideas from this field (for a general review, see [82]).

Think therefore about an Ising magnet with a surface and consider a spin in the surface layer. Let's assume we are below T_c. The surface spin has fewer neighbours than a spin in the bulk and

74

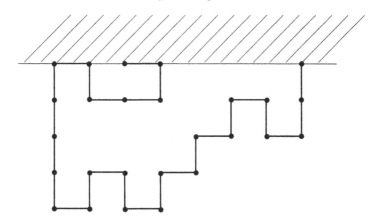

Fig. 5.1. A SAW attached to a surface.

will therefore feel less influence to align with the bulk spins. This leads to a surface magnetisation m_s which is different from the bulk value. In fact the magnetisation m will depend on the distance x from the surface and will only for 'large' x cross over to its bulk value. One can associate a 'penetration' length with this crossover. Let us introduce a simple mean field theory that describes these effects [82]. This will give the necessary insight into the behaviour of surface magnetism which will then later be useful to understand the polymer case. We consider a cubic lattice in d dimensions, and consider the semi-infinite space for which $x \geq 0$. We assume that the magnetisation depends only on the distance from the surface and therefore write it as $m(x)$, where on the lattice x runs through the non-negative integers. Within each layer (at fixed x) each spin has $z_l = 2d - 2$ neighbours. As usual we will denote by K the nearest neighbour exchange interaction (in reduced units), and will for later reference assume that the exchange interaction for spins in the surface layer can be different, let's say K_0. Finally, we introduce $\epsilon = K_0/K$. We can now write down mean field equations (in the absence of magnetic fields). These are

$$m_0 = \tanh\left(z_l K_0 m_0 + K m_1\right) \qquad (5.1)$$

$$m_n = \tanh\left(K(z_l m_n + m_{n-1} + m_{n+1})\right) \qquad (5.2)$$

Near criticality, which is always the region of interest for us, these equations can be linearised and give

$$m_0 = z_l K_0 m_0 + K m_1 \tag{5.3}$$

$$m_n = K(z_l m_n + m_{n-1} + m_{n+1}) \tag{5.4}$$

We try a solution of this difference equation of the form

$$m_n = m + A \exp(-qn) \tag{5.5}$$

where m is the bulk magnetisation and A and q are to be determined. Clearly one has that $m_s = m + A$, so we expect physically that $A < 0$ (when $K_0 = K$). Inserting (5.5) in (5.4) gives

$$[1 - (z_l + 2)K]m + A \exp(-qn) = AK \exp(-qn)(z_l + 2\cosh q)$$

The critical point is well known to be located at $(z_l + 2)K_c = 1$ (in mean field theory), while the (bulk) magnetisation vanishes near K_c with an exponent $1/2$. Consequently the first term on the left hand side will near K_c vanish as $(K - K_c)^{3/2}$. We can therefore neglect this term, since the other terms will give more dominant contributions near K_c, as we will see immediately. Expanding the resulting equation for small q leads, after some algebra, to the result (in $d = 3$ where $z_l = 4$)

$$q = \sqrt{6\frac{K_c - K}{K_c}} + O((K_c - K)^{3/2})$$

We are more interested to determine $m_s = m_0$. Inserting (5.5) into (5.3) gives after some simple algebra the following result for A

$$A = m\frac{K + z_l K \epsilon - 1}{1 - z_l K \epsilon - K \exp(-q)}$$

which after substitution in (5.5) finally gives

$$m_0 = mK\frac{1 - \exp(-q)}{1 - z_l K \epsilon - K \exp(-q)} \tag{5.6}$$

We now see that there are three possible regimes, depending on ϵ. For small ϵ we have that near K_c, $m_0 < m$ (see figure 5.2a). For large ϵ, $m_0 > m$ (figure 5.2c). At a critical value of ϵ, $m_0 = m$ (figure 5.2b). In the field of surface magnetism, one refers to these regimes as, respectively, the *ordinary*, the *surface* and the *special* regime. The result which we have derived here are of course only mean field results, but it is known from further analytical and numerical work that this picture is the correct one. It is indeed easy to understand that when the surface interaction K_0 is increased

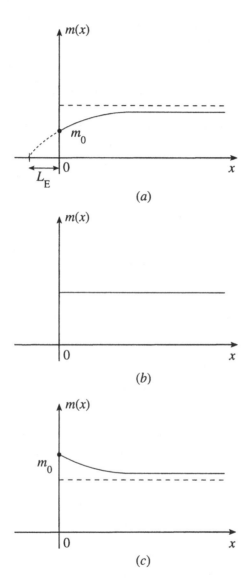

Fig. 5.2. The tree regimes of surface critical behaviour: (*a*) ordinary, (*b*) special and (*c*) surface regime. The figure shows schematically the behaviour of the magnetisation as a function of the distance to the surface in the three regimes. We also indicate the extrapolation length L_E.

there appears the possibility that the $(d-1)$-dimensional surface is ordered while the bulk is not (surface regime). An exception is the case $d = 2$ where the surface is one-dimensional and cannot order by itself unless $K_0 \to \infty$.

The results we derived also lead naturally to the introduction of the concept of *extrapolation* length L_E (see e.g. figure 5.2a). It is defined as the distance to the surface where the magnetisation (when appropriately extrapolated) goes to zero. For us it is only important to see that in the ordinary regime this extrapolation length is finite whereas it reaches infinity at the special point. We now come to a crucial point: when we study a model of surface magnetism on a coarse grained level, as is done for instance in a continuum approach, we can study the ordinary regime by including homogeneous Dirichlet boundary conditions for the magnetisation (because on this coarse grained level the extrapolation length is essentially zero). On the other hand in a continuum theory the special regime can be studied using Neumann boundary conditions. (The critical behaviour in the surface regime is of course that of the $(d-1)$-dimensional bulk system.)

~

To finish the discussion of surface magnetism, let us discuss how scaling theory has to be modified in the presence of a surface [82]. Consider therefore for the moment a finite system, e.g. a cube of side L in d dimensions. The free energy f per spin will for large L have the following expansion (we will assume that we are near a critical point, and denote by t and h the deviations from the critical point in temperature and magnetic field direction)

$$f(t,h) = f_b(t,h) + \frac{1}{L}f_s(t,h,h_s) + \ldots \qquad (5.7)$$

where f_s is a surface free energy which in principle can also depend on a surface magnetic field h_s. The latter should not influence bulk properties. The dots contain terms of higher order in $1/L$. Equation (5.7) is valid in the ordinary regime where we don't need to include a surface thermal field t_s. Such a term would describe the response of the system to an increase of the coupling for spins in the surface. From the discussion so far it is obvious that in the ordinary regime, such a term is irrelevant. In fact, there is a general argument to show that y_t^s, the scaling exponent associated with the surface thermal field, is always -1 [83] at the ordinary transition.

We are now in particular interested in the scaling behaviour of the singular part of f_s. It is of a form similar to that of the bulk free energy

$$f_s(t, h, h_s) = b^{-d+1} f_s(b^{y_t}t, b^{y_h}h, b^{y_h^s}h_s) \qquad (5.8)$$

In the exponent of b we take into account that the surface is $(d-1)$-dimensional. Bulk critical exponents cannot be modified by the presence of the surface, so y_t and y_h remain at their bulk value. We have introduced one new surface exponent y_h^s (which could in principle be the same as y_h). We can now, as usual in scaling arguments, calculate other exponents in terms of those appearing in the scaling (5.8) of the free energy. For example a surface susceptibility χ_s can be defined as

$$\chi_s(t) = \left.\frac{\partial^2 f_s}{\partial h \partial h_s}\right|_{h=h_s=0}$$

This quantity diverges near T_c with an exponent γ_s which is then immediately found to be given by

$$\gamma_s = \nu(-d + 1 + y_h + y_h^s) \qquad (5.9)$$

Exact and approximate calculations in many model systems have shown that in general $y_h^s \neq y_h$, so surface effects introduce one new critical exponent. We will now see in more detail that similar things happen for polymers attached to a surface.

5.2 The SAW near a surface

As a first step we show that polymers attached to a surface have the same connective constant as those in the bulk (general references about the surface critical behaviour of polymers are [85] and [86]). We denote by c_N^s and $c_N^{s,s}$ the number of N-step SAWs starting at the surface and ending anywhere, or ending at the surface, respectively. Similarly we denote by q_N^s the number of SAPs with at least one step in the surface. Notice that, because polygons are counted up to a translation, $q_N^s = q_N$. Furthermore, any polygon attached to a surface must have one other vertex in the surface. This leads to the following set of inequalities

$$q_N = q_N^s \leq c_N^{s,s} \leq c_N^s \leq c_N \qquad (5.10)$$

Using results from chapter 2 it is then immediately obvious that c_N^s and $c_N^{s,s}$ have the same connective constant as in the bulk. We

80   *Polymers near a surface*

therefore propose the following form for their behaviour

$$c_N^s \sim \mu^N N^{\gamma_s - 1} \qquad (5.11)$$
$$c_N^{s,s} \sim \mu^N N^{\gamma_{s,s} - 1} \qquad (5.12)$$

where γ_s and $\gamma_{s,s}$ are new surface entropic exponents. They are
not independent as can be easily seen on the basis of the following argument using scaling and the relation with the $O(n)$-model.
Consider the correlation function between a surface and a bulk
spin in the $O(n)$-model (or, a spin–spin correlation function between two surface spins). We make a high temperature expansion
for this function using the techniques of chapter 2 and then send
$n \to 0$. One finds that the boundary–bulk correlation function can
be expressed as a sum over SAWs of arbitrary length which connect the spin at the surface with the spin in the bulk. When we
sum this correlation function over the bulk spin location we get on
the one hand $\sum_N c_N^s v^N$ which according to (5.11) diverges near
$v_c = 1/\mu$ with a power $-\gamma_s$. The sum over bulk spins of the correlation function is on the other hand just proportional to the surface
susceptibility $\chi_s(t)$ where of course $t \sim |v_c - v|$. So the exponent
γ_s for the SAW is given by (5.9). Starting from a surface–surface
correlation function and following a similar reasoning one finds
the scaling relation

$$\gamma_{s,s} = \nu(-d + 1 + 2y_h^s) \qquad (5.13)$$

Using (5.13), (5.9) and the usual result $\gamma = \nu(-d + 2y_h)$ we find
the following relation (sometimes referred to as Barber's scaling
relation) [84]

$$2\gamma_s + \nu = \gamma + \gamma_{s,s}$$

In summary, if we believe in the extension of scaling to surface
critical behaviour we have to determine one more critical exponent
for SAWs near surfaces. How do we go about to determine these
exponents? There is in fact little new that we have to learn. The
same methods we encountered in previous chapters can again used.
In $d = 2$ we can use conformal invariance, the transfer matrix,
the Coulomb gas, Monte Carlo methods, and in some cases exact
solutions, whereas in $d = 3$ besides numerical methods the RG
method can be applied.

For example, one can determine the surface exponents γ_s and
$\gamma_{s,s}$ using exact enumeration techniques. A simple Monte Carlo

method to determine γ_s would be to count what fraction of a set of SAWs stays on one side of a surface. Analytical calculations are based on the Coulomb gas or on field theoretic methods where one uses the same field theories as for bulk exponents but with the boundary conditions discussed above. Often it is useful to work with an appropriate generalisation of the method of images, familiar from electrostatics. It is beyond the scope of this book to discuss these calculations (an extensive survey of this subject is given in [85]).

Let us first discuss the situation in $d = 2$. Here as can be expected one has expressions from conformal invariance [89] and Coulomb gas methods [87] that give, for instance, the surface exponent y_h^s for the $O(n)$-model for general n. The result is

$$y_h^s(n) = \frac{3}{2} - \frac{3}{4}g_R \qquad (5.14)$$

where $g_R(n)$ was already given in (3.12). Taking $n \to 0$ (or $g_R = 3/2$) and inserting this into the scaling relations leads to the following 'exact' results

$$\gamma_s = 61/64 \qquad \gamma_{s,s} = -3/16 \qquad (d = 2) \qquad (5.15)$$

We mention here that there exists a whole class of surface watermelon exponents $x_L^s(n)$ describing the decay of correlation functions which connect two points at the surface with L-lines. In terms of these exponents one obviously has $y_h^s = 1 - x_1^s$. For completeness we give the result for $x_L^s(n)$ [87]

$$x_L^s(n) = \frac{1}{4}g_R L^2 + \frac{1}{2}L(g_R - 1) \qquad (5.16)$$

The above predictions for γ_s and $\gamma_{s,s}$ have by now been well verified using exact enumerations [86] and other techniques. Most precise numerical results again come from using the transfer matrix. The technique used is similar to the one discussed in sections 2.6 and 2.7. The idea is that one starts from a semi-infinite system and considers the surface–surface correlation function. Then one performs the conformal transformation $w(z) = \frac{W}{\pi} \log z$ and uses (3.16) [45]. The semi-infinite plane is now mapped onto a semi-infinite strip of width W but now with *free* boundary conditions. Reasoning similar to that in section 2.7 then teaches us that from gaps Δ_f in the spectrum of the transfer matrix we can find surface critical exponents x^s; the relation is similar to (3.17) and becomes in this case

$$\Delta_f = \frac{\pi x^s}{W} \tag{5.17}$$

Batchelor and Suzuki [88] extended the exact solution of the hexagonal lattice $O(n)$-model discussed in section 3.4 to the case of a strip with free boundary conditions. As in that case one first maps the $O(n)$-model onto a vertex model. But in this case one has to include extra states for the vertices at the surface. Then one studies this vertex model on a strip of width W. Using the Bethe *Ansatz* technique one can determine the spectrum as a function of W. With the use of (5.17) one finally obtains the surface exponents. This work has shown that the Coulomb gas result (5.16) is indeed exact.

We conclude our discussion of the situation in $d = 2$ by mentionning another application of the principle of conformal invariance. Using a conformal mapping $w = z^{\alpha/\pi}$ one can also map half spaces into wedges with an opening angle α [89]. The transformation law (3.16) for correlation functions then allows a calculation of the way in which surface exponents for flat surfaces can be related to those for the wedge. The extreme case of a wedge of opening angle 2π describes the situation of a SAW in a 2-dimensional space with a so called excluded 'needle' [90]. At the time that conformal invariance was introduced, polymer studies played an important role in the verification of the transformation laws, since SAWs can more easily be studied in confined geometries than more standard spin systems.

In $d = 3$ there are remarkably few results for surface exponents. There exist several exact enumeration studies, some results from field theory and some not so recent Monte Carlo studies. These calculations give estimates for $\gamma_s \simeq 0.70 \pm 0.025$ [86]. It is my impression that there is room for extensive modern Monte Carlo calculations to obtain a more precise value for this exponent.

In dimensions above 4 we can again get exact exponent values from studies of random walks near a surface. Alternatively, one can extend the mean field theory discussed previously. In both cases one finds $\gamma_s = -\gamma_{s,s} = 1/2$ [82].

5.3 Polymer adsorption

We are now ready to introduce a more interesting model for a polymer near a surface, one in which the polymer interacts with the surface. Consider therefore a polymer near a surface and suppose there is an interaction between the surface and monomers in or close to the surface. To be more definite, we will consider a hypercubic lattice and take a SAW starting at some point on the surface. For each monomer on the surface the polymer gains an energy -1 (in some units). Let the SAW have N steps and V monomers in the surface (we will call the latter *contacts*). We denote by $c_N^s(V)$ the number of SAWs of N steps and V contacts. The equilibrium statistical mechanics of this model is then determined by the following (canonical) partition function (for fixed N)

$$Z_N^s(\beta) = \sum_V c_N^s(V) \exp(\beta V) \qquad (5.18)$$

Clearly, at infinite temperature we have $Z_N^s(0) = c_N^s$, whose behaviour is known from (5.11). From the discussions in this book we can expect that $Z_N^s(\beta)$ will again have an exponential form (with appropriate corrections) and one might ask whether the associated connective constant will depend on β. To answer that question, we will first show that for $N \to \infty$, the partition function behaves in a non-analytical way, i.e. a phase transition will occur when β is increased [91]. This is just the transition to be expected from the analogy with surface magnetism.

Let us define the surface 'free energy' $f_s(\beta)$ as

$$f_s(\beta) = \lim_{N \to \infty} \frac{1}{N} \log Z_N^s(\beta) \qquad (5.19)$$

Using concatenation arguments one can show that this limit exists [91]. Now clearly, $Z_N^s(\beta) \geq c_N^s(N) \exp(\beta N)$. It can be immediately realised that $c_N^s(N)$ just gives the number of SAWs on a $(d-1)$-dimensional lattice. Hence, we obtain the following lower bound for $f_s(\beta)$

$$f_s(\beta) \geq \log(\mu_{d-1}) + \beta \qquad (5.20)$$

Now let's look at the case $\beta \leq 0$; physically this corresponds to a surface interaction which is repelling. For $\beta \leq 0$, we have $Z_N^s(\beta) \leq Z_N^s(0) = c_N^s$. Furthermore, taking only the term with $V = 1$ from

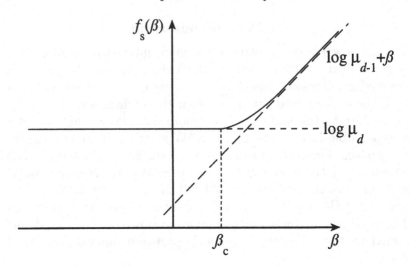

Fig. 5.3. The surface free energy as a function of β (see text).

the sum (5.18) we have $Z_N^{\mathrm{s}}(\beta) \geq c_N^{\mathrm{s}}(1) \exp \beta$. From each walk which has only the starting point in the surface we can delete the first step, move the surface one lattice unit and get a walk of $N-1$ steps with an arbitrary number of steps in this translated surface. Stated mathematically: $c_N^{\mathrm{s}}(1) = \sum_V c_{N-1}^{\mathrm{s}}(V) = Z_{N-1}^{\mathrm{s}}(0)$. Putting together the last two results, taking the logarithm and sending $N \to \infty$, we find that for $\beta \leq 0$

$$f_{\mathrm{s}}(\beta) = \log \mu_d \qquad (5.21)$$

where we have used (5.11). In figure 5.3 we show the expected behaviour of f_{s} as a function of β. The combination of the equality (5.21) and the lower bound (5.20) implies the existence of a critical value $\beta_{\mathrm{c}} \geq 0$ where f_{s} starts to increase; at that point the free energy is non-analytical, implying the existence of a phase transition. With a lot of hard work, one can in fact show that $\beta_{\mathrm{c}} > 0$ [91]. We thus recover, in an exact way, the three regimes known from surface magnetism. For $\beta < \beta_{\mathrm{c}}$ we are in the ordinary regime where the free energy is that of a polymer without interaction with the surface (in this whole regime the connective constant is therefore also equal to that in bulk). At β_{c} we have the special transition, while for $\beta > \beta_{\mathrm{c}}$ the polymer has, on sufficiently large length scales, the properties of a polymer in one dimension lower.

Indeed for $\beta \to \infty$ we have $f_s \to \log(\mu_{d-1}) + \beta$.

In a polymer language one speaks of the transition as an adsorption transition for the following reason. From (5.18) it follows that the fraction m_a of monomers in the surface is given by

$$m_a = \frac{\sum_V V c_N(V) \exp(\beta V)}{N Z_N^s(\beta)} = \frac{1}{N} \frac{\partial \log(Z_N^s(\beta))}{\partial \beta}$$

so that in the thermodynamic limit,

$$m_a = \frac{\partial f_s(\beta)}{\partial \beta} \tag{5.22}$$

From figure 5.3 it is then obvious that $m_a = 0$ in the ordinary regime, and that it starts to increase above β_c where a finite fraction of monomers is adsorbed. For $\beta \to \infty$ one reaches a regime of complete adsorption.

How does one describe our adsorption model using the $O(n)$-model? The reader should by now be familiar enough with the high T expansion of the $O(n)$-model to realise that SAWs interacting with a surface are described by an $O(n)$-model in which the coupling K is modified in the surface layer to K_0. Each contribution in the high T expansion involving a step along the surface should then get an extra weight K_0/K which has to be identified with $\exp \beta$. (Note that this gives a slightly different model from that defined by (5.18); needless to say this does not modify the phase diagram qualitatively.) In this way the adsorption model can be related to an $O(n)$-model with surface enhanced couplings. The adsorption transition at β_c is then described by studying for example continuum versions of the $O(n)$-model with Neumann boundary conditions. This will be of interest in describing the critical behaviour at the special (adsorption) point. Before we enter that study, we have to make a final point. Our proof for the existence of a surface phase transition is valid for all $d \geq 2$. In the case $d = 2$ one may wonder whether there is no contradiction in the fact that in that case the surface is one-dimensional and that according to the Mermin–Wagner theorem there can be no transitions in $d = 1$. A more detailed study shows that for $O(n)$-models with $n < 1$ the Mermin–Wagner result does not hold so that polymer models can indeed have phase transitions in $d = 1$.

Time for some exponents. Let us return to the behaviour of m_a for β slightly above its critical value. There the fraction of adsorbed monomers starts to increase; as in the case of an order parameter we can associate an exponent with this. In this case the exponent is however thermal in nature since it corresponds to the response of the model to a small change in the strength of the surface coupling (see also figure 5.3). We denote the thermal relevant eigenvalue of the RG equations at the special point by y_t^s, and introduce the *crossover* exponent which as usual is defined as $\phi_s = y_t^s/y_t$. Near the adsorption transition, the surface free energy f_s then scales as

$$f_s(t, h, h_s, t_s) = b^{-d+1} f_s(b^{y_t}t, b^{y_h}h, b^{y_h^s}h_s, b^{y_t^s}t_s) \qquad (5.23)$$

Here t_s is the surface thermal scaling field. Equation (5.23) is the extension of (5.8) to the special point. So at the special point we have to determine two new surface exponents. These are sufficient to describe the surface critical behaviour. As an example, simple scaling arguments lead from (5.23) to the scaling of m_a near β_c (remember that $t \sim 1/N$),

$$m_a = A(\beta - \beta_c)^{(1/\phi_s)-1}(1 + \ldots) \qquad (5.24)$$

In figure 5.4 we show some Monte Carlo results for m_a versus β for polymers of different length in $d = 3$ (the Monte Carlo techniques to study interacting polymers will be described in chapter 8). From data such as these one can get an estimate of ϕ_s. In $d = 3$ one finds from Monte Carlo results such as these a value of ϕ_s close to 0.5 [92]. Renormalisation group calculations give the higher value of 0.59 [93], whereas series estimates fall in between [94]. Since we are at another fixed point y_h^s is also expected to change. Indeed in $d = 3$ one has values for γ_s at the special point in the range 1.5 ± 0.05 [86].

In $d = 2$ the situation turns out to be less straightforward than in the case of the ordinary transition. From conformal invariance, Cardy obtained the result $\gamma_s = 93/64$ [89] and Burkhardt *et al.* obtained the value $\phi_s = 1/2$ [95]. These results seemed to be in agreement with data from transfer matrix studies. On the other hand, Monte Carlo studies suggested a higher value of ϕ_s [96]. A conjecture for the watermelon exponents $x_L^s(n)$ was given by Guim and Burkhardt [97]. More recently these exponents have been determined using an exact solution of the honeycomb $O(n)$-model with variable surface couplings [98]. As discussed in chapter 3 the $O(n)$-model can be transformed into a vertex model. In the case

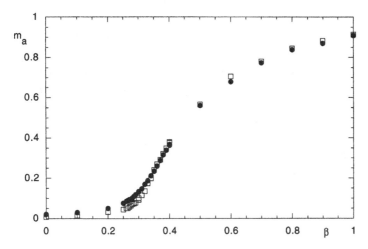

Fig. 5.4. Monte Carlo results (obtained by the pivot method for SAPs) for $m_{\rm a}$ in $d = 3$. The circles (squares) indicate results for $N = 400$ ($N = 1000$).

of surface critical behaviour, one has to include some extra vertices for the surface configurations whose weight can be modified to include a surface interaction. Using this model and the Bethe *Ansatz* technique for determining eigenvalues of the transfer matrix, the values for $\gamma_{\rm s}$ and $\phi_{\rm s}$ coming from conformal invariance have been fully confirmed. The complete set of watermelon surface exponents at the special transition for all n values have been determined and found to be in agreement with those predicted by Guim and Burkhardt. The exponents are given by

$$x_L^{\rm s}(n) = \frac{1}{4}g_{\rm R}(L+1)^2 - \frac{3}{2}(L+1) + \frac{9 - (g_{\rm R} - 1)^2}{4g_{\rm R}} \qquad (5.25)$$

Finally, the exact solution also allowed a determination of the location of the special point on the hexagonal lattice. For the polymer case which interests us here the result is [98]

$$\beta_{\rm c} = \log\sqrt{1 + \sqrt{2}} = 0.4406\ldots \qquad (5.26)$$

In summary, we see that the behaviour of a polymer near a flat surface is, especially in $d = 2$, rather well understood by now. An interesting open question is how this behaviour is modified when the surface is no longer flat. First, note that any irregularities of the surface which have some typical size should not modify the critical behaviour since using the RG one can rescale the system to a length scale on which such a surface becomes flat. So the interesting cases are those in which the irregularities of the surface have no typical length scale, i.e. when the surface becomes fractal, or self affine. Some studies have been performed for the case where both bulk and surface are fractal (extending the work described in section 3.6) [99, 100] but the case in which only the surface is fractal has so far received little attention [101, 102].

6
Percolation, spanning trees and the Potts model

In this chapter we make a side step to a subject which is not directly concerned with polymers but which is closely connected to it. It is the study of percolation. Together with problems such as SAWs and lattice animals percolation forms a subject which is sometimes called geometrical critical phenomena. As we will see in chapter 8, percolation (in $d = 2$) is closely related to the behaviour of polymers at the 'θ-point'. Moreover, percolation plays a role in the collapse of branched polymers. We will also discuss how percolation is related to a spin model (the Potts model) just as polymers are related to the $O(n)$-model. This Potts model turns out to be of importance in the description of dense polymer systems (chapter 7). The Potts model can also be used to describe the so called spanning trees. These are in turn interesting in the study of branched polymers (chapter 9).

In this book we have to limit ourselves to a discussion of those properties of percolation and Potts models which are necessary for the sequel of this book. There exists excellent reviews and books about percolation ([103–106]) and for more information we refer to these.

6.1 Percolation as a critical phenomenon

To introduce percolation, think of a regular lattice, e.g. the hypercubic lattice in d dimensions. Take a real number $0 \leq p \leq 1$ which we call the occupation probability. We occupy either the vertices (sites) or the edges (bonds) of this lattice with probability p. If one has a finite lattice of N_l bonds (or sites), the probability that

N_o of these are occupied is then given by

$$\binom{N_l}{N_o} p^{N_o} (1-p)^{N_l - N_o} \tag{6.1}$$

In percolation one asks questions concerning the *connectivity* of the graphs of occupied bonds. In the bond version of the model one says that two edges which are occupied and are incident on the same vertex are *connected*. Sets of mutually connected edges (together with the vertices on which they are incident) form *clusters*. One can then ask what the probability is that there is a cluster spanning from one end (of a finite part of) the lattice to the opposite end. In the thermodynamic limit this spanning cluster becomes the 'infinite cluster'. What is the probability that a particular bond (or site) is part of such a spanning cluster?

In the remainder of this chapter we will almost exclusively deal with bond percolation. The same results hold qualitatively for the site problem, and if one considers critical exponents, bond and site percolation are known to be in the same universality class.

A major role is played in percolation theory by the cluster numbers $n_s(p)$ and the percolation probability $P(p)$. The latter gives the probability that a particular vertex belongs to an 'infinite' cluster spanning the whole lattice. The cluster numbers $n_s(p)$ are given by $n_s(p) = p_s(p)/s$ where $p_s(p)$ is the probability that a vertex belongs to a cluster of size s (the size of a cluster will be counted by the number of edges). We also introduce $f_p(p,h)$, a quantity that plays the role of free energy for percolation

$$f_p(p,h) = \sum_s n_s(p) \exp(-sh)$$

where h plays a role similar to that of a magnetic field. One can imagine including an extra edge outside the lattice and connecting all occupied edges to this 'ghost' site. To each of these connections one then gives a weight $\exp -h$. We will not discuss the properties of $f_p(p,h)$ very much in this book, but will encounter $\kappa(p) = f_p(p,0)$ below.

It is rather obvious that $P(p)$ should be non-decreasing but the really interesting fact is that the percolation probability behaves in a non-analytical way. Indeed, one can prove that there exists a critical value p_c (also called the percolation threshold) such that

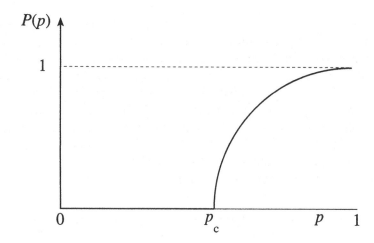

Fig. 6.1. Schematic behaviour of the percolation probability $P(p)$.

$P(p) = 0$ when $p \leq p_c$ and $P(p) > 0$ for $p > p_c$. The typical be-haviour of $P(p)$ is sketched in figure 6.1. A similarity with critical phenomena, in which $P(p)$ plays the role of order parameter, is now obvious. When $d = 1$, $p_c = 1$. One can show however that $0 < p_c < 1$ when $d \geq 2$. This proof was first given by Hammers-ley [107] and uses some results of the theory of SAWs. Indeed if a particular site (let's call it the 'origin') is in an infinite cluster it means that there must be at least one open self avoiding path from that site to 'infinity'. Now the probability of having N edges which are occupied is p^N. We can thus write that $P(p)$ is less than the probability of having a self avoiding path of N steps starting from the origin; we thus have

$$P(p) \leq p^N c_N \qquad \forall N$$

This can be rewritten as

$$P(p) \leq (p\mu)^{N+o(N)} \qquad (6.2)$$

From this equation, taking $N \to \infty$ we can conclude that $P(p) = 0$ for small enough p, in particular for $p \leq 1/\mu$ where μ is the connective constant for the lattice on which we work. Thus $1/\mu$ also gives a lower bound for p_c on any lattice. Also, as any d-dimensional hypercubic lattice can be seen as a subspace of the $(d+1)$-dimensional hypercubic lattice, we obtain at least for such lattices that $p_c(d+1) \leq p_c(d)$. So, it remains to show that $p_c(2) <$ 1. This is done by extending an argument that was originally given by Peierls [108] to prove the existence of a phase transition at finite temperature in the two-dimensional Ising model. These kind of arguments have since been called Peierls arguments. The Peierls argument for percolation is somewhat involved and requires a knowledge of some results in probability theory. It would take us too long to go into this, so we refer the reader to [104]. Combining these results then shows that the percolation threshold indeed is a point where $P(p)$ becomes non-analytic.

~

Using the idea of duality, known from graph theory, it is possible to show that on the square lattice $p_c = 1/2$ exactly. The argument goes as follows. Consider the square lattice and its dual lattice which is also square. The dual percolation process is defined in the following way; we occupy an edge of the dual lattice when it crosses an edge of the original lattice which is occupied. Since the square lattice is self dual this dual percolation process has all the properties of the original percolation problem. Then we use the famous Euler equation which relates the number of vertices v, edges s, loops l and connected components c on a graph (see figure 6.2) as

$$c = v - s + l \qquad (6.3)$$

We take a finite box of the square lattice of side L, and for a given percolation configuration in that box we can write down Euler's equation. We take the average over all percolation configurations and divide by L^2. Since for a given percolation configuration c is the number of clusters, after averaging and dividing, the left hand side of (6.3) will become equal to $\kappa(p)$. Similarly the term with the number of edges becomes equal to $2p$, whereas the term counting the number of vertices just becomes 1 (since isolated vertices are also counted as clusters (with $s = 0$)). Now, by inspecting a figure such as figure 6.2 it becomes clear that each loop on the original lattice contains a percolation cluster of unoccupied bonds on the dual lattice, apart from boundary effects. Thus, after averaging, the term in l counts the number of percolation clusters on the dual lattice, but with p replaced by $1 - p$. In this way, we deduce from

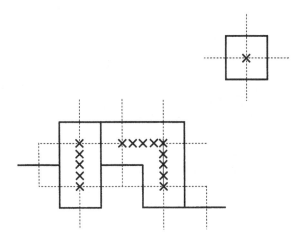

Fig. 6.2. Duality for $d = 2$ percolation on the square lattice. Occupied bonds on the square lattice are indicated as full lines, while occupied bonds on the dual lattice are shown as dotted lines. On the original square lattice, one has $l = 3$, whereas on the dual square lattice there are 4 clusters of unoccupied vertices.

equation (6.3)

$$\kappa(p) = 1 - 2p + \kappa(1 - p) \qquad (6.4)$$

We now assume that there is only one point where the percolation properties change non-analytically. Then this point has to be the percolation threshold. From (6.4) it thus follows that the unique point where κ becomes non-analytical must be at $p_c = 1/2$. Indeed, there exists by now a completely rigorous proof of this result [109]. We mention here for further reference that site percolation on the triangular lattice is also known to have $p_c = 1/2$.

The way in which P goes to zero close to p_c is described by a critical exponent β_p

$$P(p) = A_p(p - p_c)^{\beta_p}(1 + \ldots) \qquad p > p_c \qquad (6.5)$$

Near p_c one can introduce a whole set of critical exponents. In order to show further the similarity with critical phenomena, we introduce two more exponents. Firstly, one can investigate the average number of occupied edges of a (finite) cluster. This is given by the so called percolation susceptibility χ_p

$$\chi_p = \sum_s s^2 n_s$$

It diverges near p_c with an exponent γ_p. Notice that $\chi_p = \partial^2 f_p / \partial h^2$ ($p, h = 0$) which illustrates the role of f_p as generating function. Secondly, one introduces a correlation function $g(r)$ which is the probability that an edge a distance r from an occupied edge is also occupied *and* belongs to the same cluster. At criticality this correlation function decays with a power law (as usual) while for $p \neq p_c$ one can associate a correlation length ξ_p with the decay of $g(r)$. If one approaches p_c the divergence of ξ_p defines the correlation length exponent ν_p

$$\xi_p \sim |p - p_c|^{-\nu_p} \tag{6.6}$$

For these and more exponents one can set up the usual scaling relations and in the end one finds, as in the case of thermodynamic systems, that all exponents can be expressed in two exponents y_t^p and y_h^p which describe linearised RG equations near the percolation threshold. The familiar scaling relations are all recovered for percolation, e.g. one has $\beta_p = \nu_p(-d + y_h^p)$. These scalings are most elegantly written down for the free energy f_p from which the exponents of quantities such as χ_p can then be derived. We will come back to the determination of the exponents y_t^p and y_h^p when we discuss the Potts model in the next section.

$\sim$

To conclude the discussion of percolation, we describe the distribution of cluster sizes and the geometrical structure of the clusters as a function of p. In particular we will be most interested in the structure of the clusters near p_c.

Let us first introduce the idea of a *lattice animal*. As we will see these 'animals' are very important in the study of branched polymers. By lattice animal we mean a connected subgraph of a lattice (see figure 6.3). For each animal we can count the number of sites (or vertices) v, the number of bonds (or edges) s and the number of *perimeter* edges t, i.e. edges that do not belong to the animal but are incident on a site of the animal. The number of such animals which contain a given point of the lattice (the origin) is denoted by $g_s(v, t)$. If now we go back to percolation, it is easy to realise that the probability $n_s(p)$ that the origin belongs to a

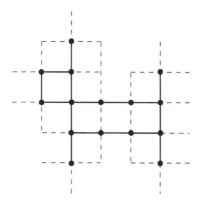

Fig. 6.3. A lattice animal with $v = 15, s = 16$ and $t = 27$.

cluster of s bonds is given by

$$n_s(p) = \sum_{v,t} g_s(v,t) p^s (1-p)^t \qquad (6.7)$$

When considering site percolation, one would have to count also the number of perimeter sites of the animal and modify (6.7) accordingly. In fact, using concatenation arguments similar to those for SAWs, one can for example prove that $g_s = \sum_{v,t} g_s(v,t)$ diverges exponentially in s. The corrections to this behaviour are again power laws and will be discussed in chapter 9. Let us call the corresponding growth constant μ_a and denote the exponent of the correction factor by θ. Thus,

$$g_s \sim \mu_a^s s^{-\theta} \qquad (6.8)$$

Now notice that when $p \to 0$ (so that $1 - p \simeq 1$) this implies that $n_s(p)$ itself grows exponentially and is essentially determined by the lattice animal growth constant. In particular, we have

$$n_s(p) \sim (\mu_a p)^s s^{-\theta} \qquad p \to 0$$

In fact, it is now known that in the whole region $p < p_c$

$$n_s(p) \sim (\lambda(p))^s s^{-\theta} \qquad p < p_c \qquad (6.9)$$

where θ does not depend on p. This all seems very similar to what we know for SAWs. When $p > p_c$ a finite fraction $P(p)$ of vertices is in the infinite cluster. What about the remaining finite clusters? It has been shown [110] that

$$n_s(p) \sim \exp\left[-\sigma(p) s^{(d-1)/d}\right] \qquad p > p_c \qquad (6.10)$$

Here $\sigma(p) > 0$ is some function of p. The corrections to the behaviour (6.10) are again power laws. The factor $(d-1)/d$ appearing in (6.10) suggests a surface effect. We will encounter similar behaviour in the study of dense polymer systems.

From several exact and numerical studies, the following picture of the structure of a 'typical' finite cluster has become clear. Below p_c they are essentially like lattice animals and have a fractal structure with large tentacles. Above p_c, finite clusters are more or less d-dimensional compact spheres.

$\sim$

We finish our little tour of the percolation process by a study of the structure of clusters at p_c. First, what about the cluster numbers $n_s(p_c)$? From extensive numerical simulations it has been verified that they follow a power law distribution at p_c

$$n_s(p) \sim s^{-\tau} \qquad p = p_c \qquad (6.11)$$

The exponent τ can be related to the exponents y_t^p and y_h^p introduced earlier. In fact one can set up a whole scaling theory for cluster numbers [111]. One can then ask the question, is there an infinite cluster exactly at p_c? At first glance the answer is no, since $P(p_c) = 0$. On the other hand one can make the following reasoning. From finite size scaling, it follows that at p_c the probability $P_L(p_c)$ that the origin belongs to a cluster that spans a box of side L goes as

$$P_L(p_c) \sim L^{-\beta_p/\nu_p}$$

or, equivalently, the average number of sites of the spanning cluster N_L in a box of side L is

$$N_L \sim L^{d-\beta_p/\nu_p} \sim L^{y_h^p} \qquad (6.12)$$

where we have used a scaling relation mentioned earlier in this section. Comparing this with (2.15) we realise that the infinite cluster at percolation is a fractal with a dimension D equal to y_h^p. This infinite cluster is often called the incipient infinite cluster (IIC) because it is there but the probability of finding an edge belonging to the cluster is zero. Yet, by studying numerically the properties of the largest cluster or spanning cluster in finite boxes of increasing size, one can learn a lot about the IIC. As the argument above shows, calculating the fractal dimension D of the IIC amounts to calculating a critical exponent. In $d = 2$, this is again possible using techniques from Coulomb gases and conformal invariance as we will see in the next section. In $d = 3$, one is mostly limited to simulations.

There is by now a great deal known about the fractal properties, both static and dynamic, of the IIC. For interested readers, we refer to [112]. For our purposes it is sufficient to introduce one more dimension (or exponent) of the IIC, the *hull* fractal dimension [113]. The hull is a subset of the set of perimeter edges. It consists of those perimeter edges from which there exists a path on the lattice to infinity that doesn't cross the infinite cluster itself. Another phrase for hull would be external perimeter. For percolation in $d = 2$ there is now a conjecture for the fractal dimension D_H of the hull of the IIC; the value is $D_\mathrm{H} = 7/4$ [114]. When we have discussed the relation between the Potts model and percolation in the next section we will indicate how the exact results for D and D_H can be obtained.

6.2 The Potts model

Percolation is connected to the q-state Potts model in a way similar to the way SAWs and the $O(n)$-model are linked. The q-state Potts model [115] is another spin model whose high temperature expansion (when $q \to 1$) gives diagrams which are percolation configurations. In the next chapter we will discuss how in turn the critical Potts model is linked to the $O(n)$-model *at low temperatures*. In the Potts model, one has at each vertex i of a regular lattice a spin variable σ_i which can be in q states ($q = 2$ corresponds to the Ising case). Interactions favour nearest neighbours which are in the same state. The effective Hamiltonian for the model is

$$H_\mathrm{P} = K \sum_{\langle i,j \rangle} (\delta_{\sigma_i,\sigma_j} - 1) \qquad (6.13)$$

In $d \geq 2$ the model has a phase transition at some $K_\mathrm{c}(q)$. In $d = 2$ and on the square lattice the location of this transition is known from duality. It is given by

$$K_\mathrm{c}(q) = \log\left(\sqrt{q} + 1\right) \qquad d = 2 \qquad (6.14)$$

It is interesting to note that for $q \leq 4$ the transition is second order, whereas for higher q values the transition becomes first order [116]. Let us perform a high T (small K) expansion of the partition function

$$Z_P(K) = \text{Tr} \exp H_P$$

This can be worked out as follows

$$Z_P(K) = \text{Tr} \prod_{\langle i,j \rangle} \exp[\delta_{\sigma_i,\sigma_j} - 1]$$

$$= \text{Tr} \prod_{\langle i,j \rangle} [\exp(-K) + \delta_{\sigma_i,\sigma_j}(1 - \exp(-K))] \quad (6.15)$$

The product in this equation can be worked out as follows. Each term in the product can be represented by a graph in which the edge between i and j of the lattice is occupied when we take the term containing $\delta_{\sigma_i,\sigma_j}$ for that edge, whereas that edge remains unoccupied if we take the other term. One particular term in the expansion of (6.15) gives a contribution

$$(1 - \exp(-K))^{N_o} \exp(-K(N_l - N_o))$$

(where we have used the notation of (6.1)) to the partition function. What remains is to work out the trace over spin states. For a cluster of occupied edges the product of Kronecker deltas ensures that within such a cluster all spins have to be in the same state. So this leads to a factor q for each cluster. Vertices that do not belong to any cluster give an extra factor of q. Thus, when we denote the number of components of the graph by C we finally get

$$Z_P(K) = \sum_{\mathcal{G}} (1 - \exp(-K))^{N_o} [\exp(-K(N_l - N_o))] q^C \quad (6.16)$$

Now, having (6.1) in mind, we see that it is natural to define $p = 1 - \exp(-K)$. In this way, for $q \to 1$, $Z_P(K)$ becomes a sum over all possible bond percolation configurations with the correct weight (of course, that sum is just equal to 1). This result holds in any dimension and any lattice on which one defines the Potts model. On the square lattice (6.14) leads for $q \to 1$ to the already known result $p_c = 1/2$.

We are now in a position similar to that already encountered for SAWs. The fact that percolation can be related to a spin model allows the use of all known techniques from critical phenomena to determine the exponents y_t^p and y_h^p. The situation is however a bit less favourable than for SAWs, since it can shown that the upper critical dimension of the Potts model is 6. This implies that results in $d = 3$ are difficult to obtain using an ϵ-expansion. This leaves simulations as the most important tool for $d = 3$. In $d = 2$ the situation is better. Here one can again, through a series of mappings onto the Coulomb gas, find exact results for exponents. We mention the relation between q and the Coulomb gas coupling g_R [40]

$$q = 2 + 2\cos\left(\frac{1}{2}\pi g_R\right) \qquad g_R \in [2, 4] \qquad (6.17)$$

Correlation functions for the Potts model can also be mapped to charge correlations in a Coulomb gas and this then leads to a determination of the percolation critical exponents: $D = y_h^p = 91/48$, $D_H = 7/4$ and $1/\nu_p = y_t^p = 3/4$. In a nice piece of work, Coniglio [117] showed that y_t^p can be interpreted as the fractal dimension of the so called *red bonds*. These are the bonds of the infinite percolation cluster which when cut disconnect the cluster.

6.3 Spanning trees

Spanning trees are well known objects in the theory of graphs [118]. Consider a connected graph (a set of vertices and edges connecting them). We mean by a spanning tree a graph that is obtained from the original graph by deleting edges in such a way that there are no loops and that a path exists between all the vertices of the set. Many properties of these graphs are known. They are important in many applications but for our purpose the most important one is that they will appear as models of branched polymers. In the next chapter we will see that they also play a role in certain models of dense polymers in $d = 2$.

We collect two important results about spanning trees: firstly, we will discuss their entropy, i.e. how one can determine the number of such trees on a given set of vertices. Secondly, we will discuss their (rather trivial) critical behaviour which will again be related to the Potts model, this time when $q \to 0$.

~

There is a well known method in graph theory to determine the number of spanning trees on a given graph. If the graph contains N vertices we have to set up an $N \times N$ matrix A whose diagonal element A_{ii} is given by the degree of the vertex i (i.e. the number of edges going out of i) and whose off-diagonal elements A_{ij} are -1 if there is an edge going from i to j and zero otherwise. This matrix is somewhat similar to a discrete version of the Laplacian on the graph. Then there is a theorem (see [118]) stating that the number of spanning trees on the graph is given by the cofactor of any element of A. On a square lattice of $L \times L$ (using periodic boundary conditions) one obtains in this way the result that the number of spanning trees S_N grows exponentially (again) in the number $N = L^2$ of sites of the lattice. More specifically one has [119]

$$S_N = (\exp 4G/\pi)^N N \qquad (6.18)$$

where $G = 1 - 1/3^2 + 1/5^2 - 1/7^2 + \ldots$ is Catalan's constant.

To see how spanning trees are related to the Potts model, consider again the high temperature expansion of the Potts model as given in (6.16). We rewrite it as

$$Z_{\mathrm{P}}(K) = \sum_{\mathcal{G}} u^{N_\circ} q^C$$

where $u = \exp(K) - 1$ and we have omitted a trivial factor $\exp(-KN_l)$. Now we insert Euler's relation (6.3) which in this case reads $C = l - N_\circ + N_l/2$ to get

$$Z_{\mathrm{P}}(K) = q^{N_l/4} \sum_{\mathcal{G}} \left[uq^{-1/2} \right]^{N_\circ} q^{l/2} q^{C/2} \qquad (6.19)$$

To continue, we will specifically work on the square lattice. At the critical point of that lattice, we learn from (6.14) that $uq^{-1/2} = 1$. So at criticality we get, apart from a trivial factor,

$$Z_{\mathrm{P}}(K) = \sum_{\mathcal{G}} q^{\frac{l+C}{2}} \qquad (6.20)$$

So far we have only been playing around with (6.16). Now let us send $q \to 0$. Then the sum in (6.20) will be dominated by those graphs in which $l + C$ is smallest. This is achieved when $C = 1$ (one component) and $l = 0$ (no loops). But these graphs are just the spanning trees of the square lattice. We thus arrive at the conclusion that spanning trees are also critical and that their exponents can again be derived from Potts exponents. This is most helpful in $d = 2$. The results for the Potts model given in the previous section then lead to the (obvious) result $D = 2$ for

Fig. 6.4. Polygon decomposition of a graph for the Potts model. A polygon is drawn around each connected component of the graph.

spanning trees. This ends our little tour of the world of spanning trees.

6.4 Polygon decomposition of the Potts model

We conclude this chapter by a brief discussion of the polygon decomposition of the Potts model [120]. This is a first step in linking the Potts model to dense polymer phases. Let us start with a particular graph in the expansion of the Potts model's partition function. Such a graph 'looks' like a bond percolation configuration, but in general its weight is different and is given by one term of (6.16). In figure 6.4 we draw one such possible graph. In the same figure we have drawn the polygon decomposition of this graph. The lattice on which this decomposition 'lives' is called the surrounding lattice. The closed loops going around the clusters of the percolation graph are the polygons. Notice that there is a polygon for each component of the percolation graph and that one has an extra polygon for each loop in the graph. Thus the number

Fig. 6.5. Graph in the expansion for the correlation function $G_3(k, l)$.

p of polygons is given by $p = C + l$. At the critical point of the square lattice Potts model the partition function (6.20) can thus be rewritten in term of polygons as

$$Z_{\mathrm{P}}(K) = \sum_{\mathcal{G}} \left[\sqrt{q}\right]^p \qquad (6.21)$$

These polygons already remind us of a set of polymers. With the polygon decomposition we can associate some watermelon-like exponents. We therefore take two sites at the centres of a face of the surrounding lattice. Take these sites a distance r apart. Then one can define watermelon-like correlation functions $G_L(r)$ in the case of the Potts model also (see figure 6.5 for an example with $L = 3$). These correlation functions have high temperature expansions in which two 'sites' a distance r apart are connected by L polygons. The correlation function $G_L(r)$ decays with an exponent $x_L(q)$ whose value for all $0 \leq q \leq 4$ is determined from Coulomb gas mappings [121]

$$x_L(q) = \frac{g_R L^2}{8} - \frac{(4 - g_R)^2}{8 g_R} \tag{6.22}$$

where $g_R(q)$ is given in (6.17). As an example, two points on the hull of the percolation cluster have to be connected by at least two polygons. For percolation $g_R = 8/3$, and as a consequence $D_H = 2 - x_2 = 7/4$.

7

Dense polymers

So far we have studied SAWs in the low fugacity regime where $z \leq \mu^{-1}$. When $z > \mu^{-1}$, the grand partition function diverges. We can still give a meaning to it by making an analytic continuation. On the other hand, we know that SAWs appear in the high temperature expansion of the $O(n)$-model and that μ^{-1} corresponds to the critical temperature of that model. The regime where $z > \mu^{-1}$ therefore corresponds to the low temperature phase of the $O(n)$-model. A spin model has of course a well defined low temperature regime and one might ask what this phase means for polymers. It is these questions which we study in the present chapter. The polymers in this phase are usually referred to as *dense* polymers. Our discussion will mostly be limited to the two-dimensional case.

7.1 The low temperature region of the $O(n)$-model

The study of polymers in the high fugacity regime can be performed in essentially two ways. The first way is to study the low temperature properties of the $O(n)$-model. This will be done using the techniques known from previous chapters; the Coulomb gas, exact solutions using the Bethe *Ansatz*, and so on. On the other hand we can study immediately the properties of walks themselves, in the regime where $z > \mu^{-1}$. Sure enough, in that region the grand partition function diverges, but the trick is to study the properties of walks in finite systems, e.g. in a finite box of volume Λ. A typical size for such a volume is $\Lambda^{1/d}$. The finite volume leads to a cutoff for the grand partition sum.

As we know from previous chapters, polymers in the dilute regime are fractals with a dimension $D < d$. This implies that for a very large polymer the density of monomers f goes to zero when we increase Λ. In the dense phase, we expect that the density of polymers is finite and, for $z \to \infty$, approaches 1, i.e. the lattice is completely occupied by monomers (although it is impossible to perform exact calculations about this behaviour on an Euclidean lattice, the exact RG method for SAWs on fractal lattices can be used to calculate $f(z)$ in the dense phase on such a lattice). In a finite system on the other hand, the fraction f of sites which are occupied by monomers is *always* non-zero.

This is then the trick to study polymers in the dense phase. Instead of working in the grand canonical ensemble we work in an ensemble in which we fix $f = N/\Lambda$ (this is nothing but a Legendre transform). To study properties of long walks we then send $\Lambda \to \infty$, $N \to \infty$ with f fixed. The dilute phase corresponds to the limit $f \to 0$. Special attention will also be paid to the case $f = 1$. A SAW with $f = 1$ is often referred to as a *Hamiltonian walk*. Closely related to Hamiltonian walks are *Eulerian trails*. A trail is a walk on a lattice which is restricted to visit each edge at most once. But vertices can be visited more than once. For example, the graphs in the high T expansion of the Ising model, or those occurring in the Nienhuis $O(n)$-model on the square lattice, are trails. A trail which visits every edge of a graph once is called Eulerian. Hamiltonian walks will be studied in the next two sections, because recent work has shown that their properties are rather subtle and show non-universal behaviour.

A first consequence of keeping f fixed is that dense polymers are no longer fractal $(D = d)$. But their other properties are far less trivial. This was discussed in a paper by Duplantier and Saleur [122], and we will follow their arguments here.

Consider first the number of self avoiding walks $c_N(f)$ of N steps which occupy a fraction f of sites of a finite volume Λ (remember that it is f that is fixed and Λ and N that vary). Duplantier and Saleur assume the following asymptotic form for $c_N(f)$

$$c_N(f) \sim [\mu_{\mathrm{D}}(f)]^N \tag{7.1}$$

This form is inspired by exact results for Hamiltonian walks which we will discuss in the next section. We will come back to the correction terms for (7.1) later. Notice that one expects a connective constant for dense walks μ_{D} which depends on f. How can we verify the conjecture (7.1) and get an idea of the function $\mu_{\mathrm{D}}(f)$?

We know that Monte Carlo methods are not very helpful in determining connective constants. Furthermore, it is hard to perform simulations in dense phases, since both growth type algorithms and algorithms based on Markov chains have high rejectance rates. Also the exact enumeration technique doesn't work very well, so the most obvious method to use is the transfer matrix. In this approach one cannot fix f but, since one works with a finite system, it is possible to explore the regime where $z > \mu^{-1}$. Following the discussion in section 3.2, we can find the average fraction of sites $\langle f_W(z) \rangle$ visited by a SAW going between two columns a distance R apart from the largest eigenvalue $\lambda_W(z)$. Indeed, from the expressions (3.8) and (3.11), we easily find that

$$\langle f_W(z) \rangle = \frac{1}{W} \frac{\partial \log \lambda_W(z)}{\partial \log z} \qquad (7.2)$$

On the other hand, the number $c_{W,N}$ of N-step walks in the strip of width W has the behaviour (we only consider the dominant term for $N \to \infty$)

$$c_{W,N} \sim [\mu_{D,W}]^N \qquad (7.3)$$

This equation (multiplied by an appropriate scaling function in r) can be inserted in (3.8). In a saddle point approximation, the sum in (3.8) will be dominated by that term for which N is approximately $W R \langle f_W(z) \rangle$, so that

$$G_{0,W}(k,l) \approx [\mu_{D,W} z]^{W R \langle f_W(z) \rangle} \qquad (7.4)$$

Comparing with (3.11) we get

$$\mu_{D,W}(z) = \frac{[\lambda_W(z)]^{1/(W \langle f_W(z) \rangle)}}{z} \qquad (7.5)$$

Combining (7.2) and (7.5) allows us to determine $\mu_{D,W}$ as a function of $\langle f_W \rangle$. In the thermodynamic limit $W \to \infty$ this function should approach $\mu_D(f)$. Transfer matrix calculations of this kind were performed by Duplantier and Saleur [122] who found that their numerical results for $\mu_D(f)$ were indeed consistent with the known value for $f \to 0$ (SAW limit). For $f \to 1$ they obtained an estimate for the connective constant of Hamiltonian walks on the square lattice $\mu_H = 1.473$.

$$\sim$$

Secondly, we turn to the critical behaviour of dense polymers. Clearly, in $d = 2$, we have $\nu = 1/2$ in this phase. More interesting is to determine the watermelon exponents for dense polymers. As argued in the introduction, dense polymers are described by a low temperature $O(n)$-model. For $n \neq 0$ this represents a collection of loops on the lattice with the usual weight $v^N N^l$, but now the total number of vertices visited by all loops fills a finite fraction of the lattice. There exists yet another way to interpret the low temperature $O(n)$-model. One can derive the loop gas also from a low temperature expansion for the so called n-component corner cubic model [123]. In that model one has a spin variable at each vertex of a lattice that can point to the $2n$ corners of an n-dimensional hypercube. The interaction between the spins is the usual (anti) ferromagnetic nearest neighbour interaction. We don't go into the details of that model, but for $n = 1$ it is again just an Ising model. A low temperature expansion of the Ising model on the triangular lattice gives graphs which consists of loops on the dual, hexagonal, lattice with a weight $v = \exp{(-2K)}$ per broken bond. In this way we recover the diagrams of the $n = 1$ loop gas (this is just duality). When we consider a ferromagnetic Ising model, we have $v \leq 1$, whereas the regime $v \geq 1$ corresponds to the antiferromagnetic case. In a similar way the low temperature expansion of the n-component corner cubic model yields the diagrams of the loop gas for general n, which in the high v region correspond again to the cubic model with antiferromagnetic interactions. This then gives us another way to interpret the low temperature phase of the $O(n)$-model.

~

We now return to a study of the critical behaviour of the low temperature $O(n)$-model. In general a spin model below its critical point does not have long range correlations, and is thus not critical. On the other hand, remember the situation for the XY-model in $d = 2$ which has critical behaviour for all temperatures below a certain T_c, with exponents that vary continuously with T [124]. In fact, the XY-model is nothing but the $O(n = 2)$-model. So it comes as no surprise that for general n the low temperature phase of the $O(n)$-model can also be critical. This was first realised by Nienhuis from his mapping of the hexagonal $O(n)$-model onto the Coulomb gas [37]. In figure 7.1 we draw a more complete phase diagram of the $O(n)$-model on the hexagonal lattice. The upper branch shows the critical temperature as a function of n as given

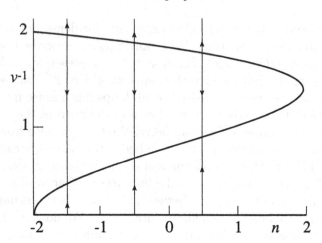

Fig. 7.1. Phase diagram of the hexagonal lattice $O(n)$-model. The upper (lower) branch is given by equation (3.7) ((7.6)). The line at $v^{-1} = 0$ represents the fully packed regime (section 7.3). The RG flow is also indicated.

in (3.7). The lower branch is given by [37]

$$v_D(n) = \frac{1}{\left[2 - \sqrt{2 - n}\right]^{1/2}} \qquad (7.6)$$

The RG flow indicates that the whole low temperature regime is determined by this fixed point. Along this branch the $O(n)$-model renormalises again onto the Coulomb gas vacuum. The relation between n and g_R is still given by (3.4) but now we have $0 < g_R < 1$. The result (3.5) still holds for the watermelon exponents, but of course with g_R modified. In particular, for the polymer case we have [125]

$$x_L^D = \frac{1}{16}(L^2 - 4) \qquad (7.7)$$

As an example, $x_2 = 0$, implying $\nu = 1/2$ as already discussed above.

As shown in figure 7.1 we see that at $n = 2$ there is a marginal thermal eigenvalue where the two critical branches coalesce. This marginal eigenvalue is related to the continuously varying exponents of the XY-model. For $n < 2$ the low temperature regime is critical but the exponents do not vary continuously. For $n < 1$, we have $x_1^D(n) < 0$, implying that correlation functions increase with distance. But that is not really a problem; in the polymer case it represents the fact that the two ends of the polymer repel each other in the dense phase.

For $n \to 0$, (7.7) predicts in particular $x_1^D = -3/16$. This value is indeed consistent with a numerical determination of x_1^D from the first gap in the spectrum of the transfer matrix [122]. From this value of x_1^D, the usual scaling relations predict the γ-exponent for the dense phase to be $\gamma_D = 19/16$. Also observe that since for a dense phase we have $\nu = 1/2$, the usual scaling relation gives for the two-dimensional case, $\alpha = 1$. But, it was argued by Duplantier and Saleur [122] that these exponents play a role different from that for usual polymers. Their argument is as follows. We start from the scaling for $c_N(r)$ given in (2.36). This scaling is derived from quite general arguments and should also hold in the dense phase. Inserting the values for γ and ν, we get

$$c_N(r) \approx q_N r^{3/8} \hat{H}(\frac{r}{N^{1/2}}) \tag{7.8}$$

When we integrate this equation over r we get a scaling law for the ratio of open SAWs to SAPs

$$\frac{c_N}{q_N} \sim N^{19/16} \qquad \text{in the dense phase} \tag{7.9}$$

This is the place where the exponent γ_D shows up. It is however impossible to predict the form of c_N itself. Indeed, remember that the spin–spin correlation function can be expressed in terms of SAWs, equation (2.25). Now in the dense phase, and in the thermodynamic limit, this sum diverges. In a finite system (of volume Λ) we expect the most important term of the sum to be given by

$$G_{0,\Lambda}(k,l) \sim [z\mu_D(f)]^{\Lambda f}$$

(compare with (7.4)). We have not written down the correction terms. These could describe the effect of boundaries, for example. A boundary effect can be expected since a dense polymer in a finite volume feels very much the boundaries of the system in which it is. Since the precise form of $G_{0,\Lambda}$ is not known we are unable to predict the precise form of c_N. Now, for the case of dense polymers on the Manhattan lattice it is known (see next section) that the simple form (2.6) which we have always used for c_N fails, precisely through boundary effects. In fact, the results on the Manhattan lattice suggest that the appropriate form for c_N in a dense phase is

$$c_N \sim \mu^N \mu_1^{N^\sigma} N^{\gamma-1} \tag{7.10}$$

where $\sigma = (d-1)/d$, i.e. the term in μ_1 represents a surface correction. Moreover, as the example of the Hamiltonian walks on the Manhattan lattice shows, γ in (7.10) may depend on the boundary conditions, the shape of the volume Λ, etc.

In summary then, Duplantier and Saleur predict that the exponent $\gamma_D = 19/16$ can only be observed in the ratio c_N/q_N. The number of SAWs c_N itself is conjectured to be of the form (7.10) with a non-universal γ.

$$\sim$$

More recently, the values of the watermelon exponents in the low temperature phase of the $O(n)$-model have been confirmed by an exact calculation on the hexagonal lattice [39]. In the square lattice $O(n)$-model of section 3.4 the same exponents can be recovered along branch 2 [58]. So at this moment there is little discussion about the precise value of these exponents.

Whether (7.9) and (7.10) hold in the dense phase is much less clear. One of the main reasons is that it is so difficult to do any simulations in these regimes. Grassberger and Hegger [126] performed Monte Carlo calculations in these regime using a new simulation method for polymers, introduced by Grassberger [127]. They remark that it is not clear whether the method works well for the dense phase. Nevertheless, they find evidence for the relation (7.9). As we will see in the next chapter, polymers in the 'collapsed phase' are similar to dense polymers. In that regime, there exist other arguments that the form (7.10) holds. In all, however, the interesting predictions (7.9) and (7.10) are still not very well verified.

In the next section we turn to a simplified model for dense polymers for which much more exact results can be obtained (i.e. not only exponents). These results will further support the picture presented in this section.

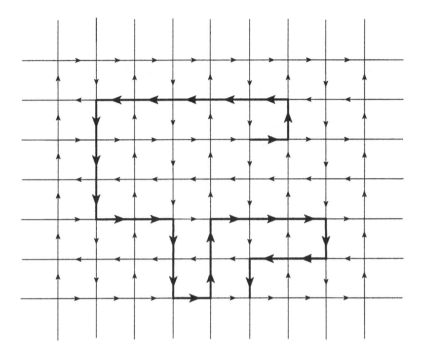

Fig. 7.2. A SAW on a Manhattan lattice.

7.2 Hamiltonian walks on the Manhattan lattice

Figure 7.2 shows a polymer on the so called Manhattan lattice. This is a directed lattice on which walks have to follow the arrows on the edges, which are alternately up/down and left/right.

We begin by mentioning that numerical work has shown that the critical exponents ν and γ for an ordinary SAW (i.e. the non-Hamiltonian case) on Manhattan are the same as those for a SAW on any undirected two-dimensional lattice [128]. So it seems that the directedness of the lattice is irrelevant in this situation. This will turn out to be no longer true for polymers at the θ-point (chapter 8). In this section we will only interested in the behaviour of Hamiltonian walks on the Manhattan lattice.

We will discuss two topics. Firstly we will show how one can find the number of Hamiltonian walks on the Manhattan lattice using a relation with spanning trees. Secondly, we will discuss the situation of a melt of Hamiltonian walks. This is the case

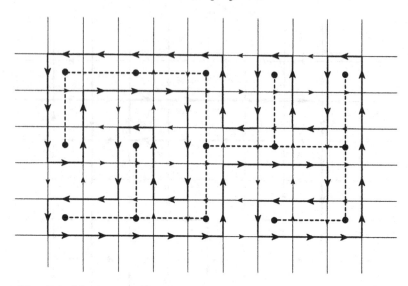

Fig. 7.3. Mapping a Hamiltonian walk on the Manhattan lattice
onto a spanning tree (thick dotted lines).

that we have an arbitrary number of walkers on the lattice which
(together) visit all vertices of the lattice.

$\sim$

By considering some examples of SAWs on the Manhattan lattice it becomes clear that such walks can never become trapped
and stop when they arrive at a nearest neighbour of the starting
site. This holds in particular for Hamiltonian walks. Let h_N be
the number of such Hamiltonian walks on the Manhattan lattice.
In figure 7.3 we show how each Hamiltonian walk on a Manhattan lattice with periodic boundary conditions can be related to
a spanning tree on the square lattice [119]. The vertices of this
square lattice are located in the centres of those plaquettes of the
Manhattan lattice where the arrows circle counterclockwise. From
this equivalence, one finds that h_N is proportional to the number
of spanning trees on the square lattice (the proportionality factor
is 8; a factor of four comes from the number of starting points of
the open walk per root of the spanning tree, and a factor of two
comes from the number of orientations of the walk). In section 6.3
we discussed a general recipe which allows the calculation of the
number of spanning trees on a lattice in terms of a determinant,

leading to the result (6.18) for the square lattice with periodic boundary conditions. Thus we immediately get the desired result for h_N

$$h_N \sim [\exp(G/\pi)]^N \qquad (7.11)$$

on a square of side L with $N = L^2$. Barber [129] was able to calculate the leading correction factor to this form and found

$$h_N \sim [\exp(G/\pi)]^N N \ldots \qquad (7.12)$$

Here the ... represent further correction terms which have a weaker N-dependence. Equation (7.12) shows that h_N is of the form familiar from SAWs. If we take into account the fact that in the relation with spanning trees, the starting point of the Hamiltonian walk is not fixed, we find that in the number of Hamiltonian walks *per lattice site*, the factor N drops out, and so we arrive at a γ-exponent for Hamiltonian walks, $\gamma_H = 1$. But this is not the complete story. In a seminal work on Hamiltonian walks on the Manhattan lattice, Duplantier and David [130] were able to determine also the number of h_N^0 of Hamiltonian walks on a Manhattan lattice with open boundary conditions, again using a relation with spanning trees. By making an expansion for large N they found the result (for a square box)

$$h_N^0 \sim [\exp(G/\pi)]^N \left[1 + \sqrt{2}\right]^{-\sqrt{N}} N^{3/4} \ldots \qquad (7.13)$$

This is a very interesting result. It shows two important points. Firstly notice that h_N^0 is indeed of the form (7.10), i.e. it contains a surface correction factor. Secondly, we see that the exponent γ_H is no longer universal; for open boundary conditions, γ_H takes on another value $\gamma_H^0 = 3/4$. These two results provided the main motivation for the conjectures of Duplantier and Saleur for dense polymers, discussed in the previous section.

~

Let us now return to our discussion of dense polymers. Hamiltonian walks represent the extreme case that $f = 1$. So, by studying Hamiltonian walks on the Manhattan lattice, we have a model in which we can study dense polymers in the limit $f \to 1$ and we can verify whether some of the exponents of the low temperature $O(n)$-model can be found in this simplified case. Stated otherwise: if the exponents of Hamiltonian walks are the same as those of the low temperature $O(n)$-model, we could say that the directedness of the lattice is irrelevant. As we will see below the situation is more complicated than this.

Our goal is now to determine the watermelon exponents of the Manhattan walk [131]. In fact, our starting point will be far from Hamiltonian walks, but will be the polygon decomposition of the Potts model, discussed in section 6.4. There we discussed how the partition function $Z_P(K)$ *at criticality* can be written as a sum over graphs consisting of p polygons covering the surrounding lattice and with a weight $(\sqrt{q})^p$ (see figure 6.4 and (6.21)). Each of the polygons in this graph is in fact a trail. But it is a trail of a special kind, namely one that at each step either turns to the right or to the left. Such a trail is sometimes referred to as an L-trail. As shown in figure 7.4 there is a one-to-one correspondence between any L-trail and a walk on the Manhattan lattice. So, we have still another way of viewing the partition function of the Potts model at criticality. It consists of all possible graphs on the Manhattan lattice in which we have a set of p SAWs visiting all vertices of the lattice. This can be seen as a simple model for a dense melt of polymers. When we send $q \to 0$ the critical partition function of the Potts model will just be described by one polygon covering the whole lattice. Using the connections established above this can also be described as one Hamiltonian walk on the Manhattan lattice, which in turn corresponds to a spanning tree. In this way, we recover the known fact that spanning trees are described by the limit of the Potts model as $q \to 0$.

When one has a set of walks which together visit every vertex of a lattice one also speaks of a fully packed model. These fully packed models reduce to Hamiltonian walks when just one SAW fills the whole lattice, i.e. for $q \to 0$. In this way, we can say that the fully packed model on the Manhattan lattice is described by the q-state Potts model. This relation then allows also a determination of all the watermelon exponents for the fully packed model, and in particular for the Hamiltonian walks on Manhattan. These are just the watermelon exponents for the Potts model described in chapter 6.

At this point there seem to be two candidates to describe dense polymers; the first one is the low temperature $O(n)$-model described in the previous section, the second is the critical Potts model which describes the Hamiltonian limit $f = 1$ on the Man-

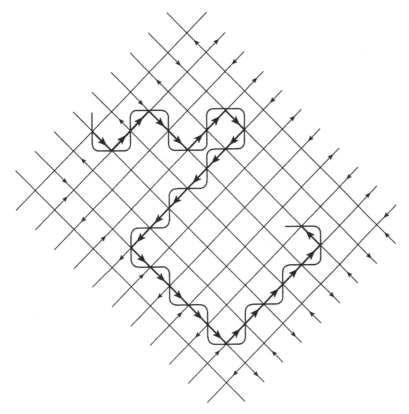

Fig. 7.4. Mapping an L-trail into a walk on the Manhattan lattice. For clarity, the arrows were not drawn on all lattice edges.

hattan lattice. Might these two be somehow related? The answer is yes, but the full details have only become clear in recent years. To discuss the relation between the two models, first notice that when one puts $\sqrt{q} = n$ in (6.21) the polygon decomposition of the Potts model resembles the loop gas representation of the $O(n)$-model. We can then start to compare the watermelon exponents of the two models (this was first done in [131]). Take a watermelon configuration with L legs going between two sites k and l in a graph of the Potts model (see e.g. figure 6.5). Then in the polygon or walk interpretation this corresponds to a situation in which there are $2L$ SAWs going between k and l. It is then possible to check using the expressions (3.4), (3.5), (6.17) and (6.22) the following equivalence: the exponent $x_L(q)$ for the q-state critical Potts model is the same as the exponent $x_{2L}(n)$ in the dense phase of the $O(n)$-model if we take $n = \sqrt{q}$. This holds for all q. In the context of the present book, we are of course most interested in the limit $q \to 0$, $n \to 0$. So, at first sight, it seems that the low temperature $O(n)$-model and the critical Potts model do indeed describe the same behaviour and that we thus have one universality class for dense polymers, independent of whether one is in the Hamiltonian limit or at some smaller density. Furthermore, such a result would imply that the directedness of the Manhattan lattice is irrelevant.

There is however a difference; the exponents $x_L(n)$ of the dense $O(n)$ model for L odd cannot be obtained from the Potts model equivalence. This situation has been further clarified in the work of Batchelor and his coworkers [132]. These authors work with the square lattice $O(n)$-model introduced in section 3.5. If in that model one takes $v = 0$ the resulting graphs contain only L-trails, or equivalently a set of SAWs on the Manhattan lattice. This is the so called branch 0 of the square lattice $O(n)$-model. For $n \to 0$, it describes Hamiltonian walks on the Manhattan lattice [132]. The resulting model is, for $n \to 0$, again solvable by Bethe *Ansatz*. When one puts the model on a strip, and calculates the gaps in the spectrum of the transfer matrix, one finds the exponents also present in the Potts model, i.e. the exponents x_{2L} of the dense $O(n = 0)$-model (chapter 3). However, the low temperature $O(n = 0)$ exponents for a watermelon with an odd number of legs are not present in the spectrum. The fact that these exponents do not occur is related to the fact that one cannot study walks on a Manhattan lattice of an odd width.

As a result of this, the exponent $x_1 = -3/16$ of the low temperature $O(n)$-model (which led to the prediction $\gamma_D = 19/16$) is not present for the Hamiltonian walks on the Manhattan lattice. The magnetic exponent on that lattice is then given by the next exponent, i.e. by $x_2 = 0$, of the dense $O(n)$-model. This leads to the result $\gamma_{H,M} = 1$ for the γ-exponent for Hamiltonian walks on the Manhattan lattice. As discussed in the previous section, this exponent could show up in the ratio c_N/q_N, but should be distinguished from the exponents γ_H and γ_H^0 appearing in (7.12), respectively (7.13).

To summarise, in this section we have discussed exact results for Hamiltonian walks on the Manhattan lattice. We have seen that the set of critical exponents is a subset of the exponents of the dense $O(n)$-model, leading, for example, to a non-universal value of γ for dense systems. Furthermore we have found evidence for the form (7.10) for c_N in the dense case.

This discussion now leads to the following question. Is the non-universality caused by the directedness of the Manhattan lattice or is it due to the Hamiltonian limit? This question is investigated in the next section.

7.3 Fully packed $O(n)$-models

In the next chapter we will see that polymers in the collapsed phase have been conjectured to be described by the dense phase of the $O(n)$-model. But which dense phase, as we saw in the previous sections that there is a subtle non-universality between the dense phase on the hexagonal lattice and the Hamiltonian walks on a Manhattan lattice? This question has, among other reasons, led to a more thorough investigation of dense polymer phases in recent years.

Let us begin by discussing the situation on the hexagonal lattice. Consider again the partition function of the $O(n)$-model and its phase diagram as given in figure 7.1. Since on the hexagonal lattice the number N of bonds visited in any graph is always even, the partition function of the loop gas has a $(v \leftrightarrow -v)$-symmetry. This leads to the idea that the line $v^{-1} = 0$ in figure 7.2 cannot be attracted to the low temperature fixed point and represents in itself a new phase, with still different exponents. If we see the loop gas as the low temperature expansion of a cubic model, as mentioned in section 7.1, the case $v^{-1} = 0$ corresponds to the ground state of the antiferromagnetic n-component cubic model.

This fully packed hexagonal model was consequently studied by Blöte and Nienhuis [133] using a transfer matrix approach in which the central charge of the model and its scaling dimensions were determined numerically. Somewhat later this fully packed model was solved by Batchelor *et al.* [134] using a Bethe *Ansatz* technique (see also [135]). These calculations do indeed confirm that the v^{-1} point represents a new universality class. But that class is closely related to that of the low temperature $O(n)$-model. In fact the central charge in the fully packed case is just 1 higher than in the dense phase at the same value of n. For the case of polymers this leads to a central charge of -1. The watermelon exponent $x_1(n)$ is given by $1 - 1/(2g_R(n))$ where g_R is still related to n via (3.4), but in this case with $g_R \in [1/2, 1]$. For the polymer case this leads to a value of $\gamma_{HFP} = 1$ for the hexagonal fully packed (HFP) case. This is the same as for the Manhattan lattice. Some of the watermelon exponents of the two models coincide but not all of them. The work of Batchelor *et al.* [134] has also led to an exact determination of the connective constant for Hamiltonian walks on the hexagonal lattice. The value is $\sqrt{(3\sqrt{3}/4)} = 1.13975\ldots$.

Finally, what is the critical behaviour of a fully packed model on the square lattice? Here, there is so far no exact solution known. Very recently this model was studied numerically using transfer matrix methods by Batchelor, Blöte, Nienhuis and Yung [136]. To confuse the situation even more, in this case the value of γ seems again different. For the case $n = 0$ these authors give the estimate $\gamma_{SFP} = 1.0444 \pm 0.0001$. In the same study a precise numerical estimate of the connective constant for Hamiltonian walks on the square lattice was determined, whose value is 1.47279. Preliminary transfer matrix calculations by the same authors, however, show that on the triangular lattice, the fully packed model seems to be described by the low temperature $O(n)$-model.

In conclusion, then, we have seen that in $d = 2$ the dense polymer phase has been the focus of a lot of work in recent years, using mainly numerical transfer matrix calculations and exact solutions. The more traditional polymer methods, based on exact enumeration or Monte Carlo simulations, fail in this regime. These studies have shown that, as can be expected, dense polymers are always compact objects with $D = 1/\nu = 2$. The γ-exponent on the other hand seems very sensitive both to the lattice under study and to the boundary conditions involved. Further in this book we will encounter other situations where the ν-exponent turns out to be very robust whereas the γ-exponent is strongly dependent on, for instance, restrictions of topological nature.

8

Self interacting polymers

So far we have discussed the properties of polymers in a good
solvent. In real situations, this is the case at 'high' temperatures.
When the temperature is lowered the quality of the solvent decreases. As a consequence excluded volume effects become smaller
and the monomer–monomer interaction becomes more important
then the monomer–solvent interaction. This self interaction, which
is attractive, leads at low enough temperatures to a collapse of the
polymer. In fact, there is a phase transition associated with this
collapse. The point at which the collapse occurs is usually referred
to as the θ-point. The temperature associated with this point will
be denoted as T_θ. At temperatures below the θ-point, a polymer
assumes a compact shape.

8.1 Polymer collapse as a tricritical phenomenon

In the study of polymers using lattice models, one introduces the
following SAW model to study the collapse transition (figure 8.1).
We begin by defining a self contact as a pair of monomers which
are on nearest neighbour lattice sites but which are not joined by
an edge of the SAW. To each such self contact we associate an
attractive energy -1 (in some arbitrary units). When we study
this model in an equilibrium ensemble each SAW is weighted by a
factor $\exp(\beta I)$ where I is the number of self contacts. Thus, the
appropriate partition function for N-step SAWs is

$$Z_N^c(\beta) = \sum_I c_N(I) \exp(\beta I) \tag{8.1}$$

where $c_N(I)$ is the number of N-step SAWs which have I self
contacts. With this partition function one can associate a free

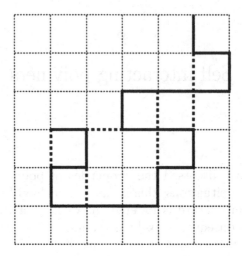

Fig. 8.1. Model for polymer collapse. The dotted lines indicate the self contacts.

energy $f_c(\beta)$

$$f_c(\beta) = \lim_{N \to \infty} \frac{1}{N} \log Z_N^c(\beta) \qquad (8.2)$$

$(\beta = 1/T)$. There is only a proof that the limit in (8.2) exists for the physically less interesting regime where $\beta \le 0$ [137]. In that case one also knows that a corresponding free energy for SAPs exists (for all temperatures) and has the same value as that for walks. In general it is not known how to prove exactly that the free energy (8.2) has a singularity, as could be done in the case of the surface free energy $f_s(\beta)$. We will assume that the free energy in (8.2) exists for all β and that it defines a connective constant $\mu_c(\beta)$ and a γ_c exponent

$$Z_N^c(\beta) \sim [\mu_c(\beta)]^N N^{\gamma_c - 1} \qquad (8.3)$$

Whereas μ_c can be expected to vary continuously with β, we expect γ_c to take on three distinct values. At high temperatures γ_c is equal to the value for a non-interacting SAW. At the θ-point it takes on another value γ_θ, and at low temperatures it will take on a third value. As we will argue in section 5 of this chapter, in the collapsed phase polymers are expected to be similar to dense polymers, so that the non-universal aspects of γ discussed in the previous chapter will show up. Similar behaviour is shown by the exponent ν. Above the θ-point it takes on its non-interacting SAW value. In the low temperature phase, the polymer collapses into a compact configuration with an exponent $\nu = 1/d$. Finally, precisely at the θ-point this exponent takes on a third value ν_θ. At that point there is also a crossover exponent ϕ_θ. In the present chapter we will discuss what is known about these exponents at the θ-point and about the behaviour of the polymer in the low temperature phase.

$$\sim$$

Let us however first turn to the grand canonical description of self interacting polymers. We know from the previous chapters that long polymers constitute a critical system. In fact their criticality is that of the $O(n = 0)$-model. From (8.3), we can understand that polymers are critical for all temperatures and that their grand partition function $\mathcal{Z}^c(z, \beta) = \sum_N z^N Z_N^c(\beta)$ becomes non-analytical at $\mu_c(\beta)$. In figure 8.2 we sketch the phase diagram of the self interacting polymer using RG flows (for an example of a real space RG calculation, see [138]). For $T > T_\theta$, the critical point flows, under the RG transformation, to a fixed point at $\beta = 0$. In this regime the whole critical behaviour is therefore that of $\beta = 0$. When on the other hand $T \leq T_\theta$, the RG flow is to a fixed point at $T = 0$ where the thermal eigenvalue is equal to d (because the walk is compact). Such an eigenvalue is however characteristic of *discontinuity fixed point*, which describes a first order transition. This implies for instance, that the density of monomers makes a finite jump at $z_c(\beta)$ when we are in this low temperature phase. This is a consequence of the fact that the density is related to the derivative of $\mathcal{Z}^c(z, \beta)$ with respect to z, which is exactly the field having the relevant eigenvalue d. Then standard RG arguments predict a first order jump in the monomer density. Since the θ-point itself thus separates a line of second order transitions from a line of first order ones, it has all the properties of a tricritical point (this was first argued in [139]). Now, we come to an important point. It is known that for tricritical phenomena, the upper critical dimension is 3. This leads to the prediction that the θ-point for $d = 3$ is described by mean field exponents $\nu = 1/2$, $\gamma = 1$. But, because one is exactly at the upper critical dimension the leading power law behaviour of most quantities has strong logarithmic corrections.

Self interacting polymers

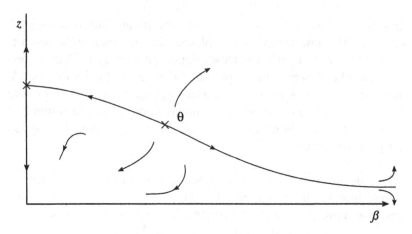

Fig. 8.2. Schematic RG flow for the polymer with self interactions.

So, in summary, in $d = 3$ the critical exponents at the θ-point are supposed to be mean field like with strong logarithmic corrections. This leaves only the exponents in $d = 2$ undetermined.

In the next section we will discuss a model for interacting polymers in $d = 2$ which differs slightly from that of (8.1) but which can be solved exactly because it can be related to the low temperature $O(n)$-model.

In section 8.3 we will describe how some of the numerical and analytical approximate techniques known from chapter 2 and 3 have to be modified to study interacting polymers.

8.2 A model in $d = 2$

The critical exponents at the θ-point in $d = 2$ are now known thanks to a lot of work starting around 1986. At that period, Coniglio *et al.* [140] introduced a model for an interacting polymer which is slightly different from the lattice model described in the previous section. Since it was not *a priori* clear that this model was in the same universality class as the usual model, it was called the θ'-model. A more extended discussion of Coniglio's model was later given by Duplantier and Saleur [141], and we will discuss their work here.

Consider therefore a SAP on the hexagonal lattice. In the model,

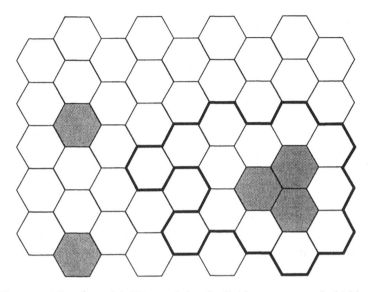

Fig. 8.3. The θ'-model. Edges of the shaded hexagons are forbidden to the SAW.

hexagons of the lattice are removed with a probability $1-p$ and as a consequence the SAP cannot visit the edges of such a hexagon (figure 8.3). Since the centres of the hexagons form a triangular lattice, this defines a percolation process on the dual, triangular lattice. Remember that this percolation problem is critical at $p_c = 1/2$. Now, keep the SAP fixed and sum over all configurations of the lattice in which that particular SAP can occur (i.e. consider the randomness as annealed). A necessary and sufficient condition for a configuration to contribute to this sum is that all hexagons whose edges are visited by the walk should be allowed for the SAP. If the SAP has N steps and visits H hexagons, the weight of that particular SAP after performing the average over hexagon configurations becomes p^H. If we denote by $q_N(H)$ the number of N-step SAPs that visit H hexagons we can define a partition function,

$$Z_N^{\theta'}(p) = \sum_H q_N(H) p^H \qquad (8.4)$$

The corresponding grand partition function is then given by

$$\mathcal{Z}^{\theta'}(z,p) = \sum_N z^N Z_N^{\theta'}(p) \tag{8.5}$$

In a similar way we can define a model for SAWs, but as we will see below, in the present case, it is more convenient to work with polygons.

We can now explain why this model is a model for a self interacting polymer (figure 8.4). The number of H of hexagons visited by the walk reaches its maximum value for a 'straight' walk for which $H = N + 1$. In other cases, some hexagons are visited twice (let their number be N_2)(figure 8.4a, at hexagon 1) or three times (N_3)(figure 8.4a, hexagon 2). In this way we obtain $H = N + 1 - N_2 - 2N_3$. Now, from figure 8.4 it is obvious that $N_2 + 2N_3$ is at least equal to the number of self contacts. In fact, we can write $N_2 + 2N_3 = I + I'$ where I' counts the number of a subset of next nearest neighbour contacts (figure 8.4b). Thus, we can rewrite (8.5) as a model for an interacting polymer

$$\mathcal{Z}^{\theta'}(z,\beta) = p \sum_N \sum_{I,I'} q_N(I,I')(zp)^N \exp\left(\beta(I+I')\right)$$

where $\beta = -\log p$. Because this model includes the number of a certain type of next nearest neighbour interactions, it is slightly different from the θ-model of the previous section, but because of universality, we expect both models to have the same critical behaviour.

How do we go about to determine the critical exponents of the model? We look back at figure 8.3 and trace out the boundaries of the sets of forbidden hexagons. In this way the configuration of hexagons and one SAP is mapped onto a gas of loops. The boundaries of forbidden hexagons in fact represent the hulls of the percolation clusters on the triangular lattice. The weight of a particular configuration of the loop gas is $z^N p^{H_a}(1-p)^{H_f}$ (where obviously H_a and H_f stand, respectively, for the number of hexagons which are allowed and forbidden to the walk). The idea is now to relate this particular loop gas to the loop gas representation of the $O(n)$ model. This can be achieved by taking $z = 1, p = 1/2$, because in that case the weight of any configuration becomes $(1/2)^{H_t}$ where $H_t = H_a + H_f$ is the total number of hexagons in the lattice. In that case we have a loop gas in which *all* configurations have the

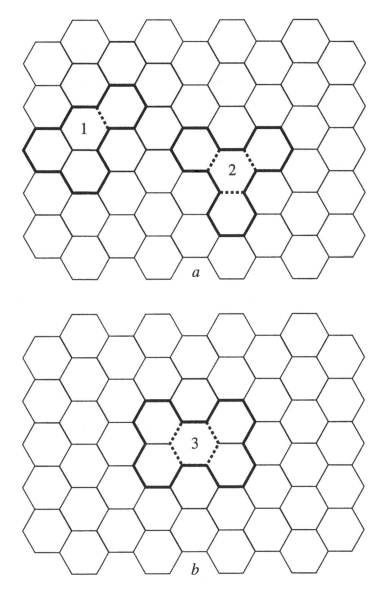

Fig. 8.4. Interactions in the θ'-model (see text).

same weight. Looking back at (2.21) this is Nienhuis's loop model at $K = n = 1$. Since at $n = 1$, $v_c < 1$ (see (3.7)), we see that the θ'-model at $z = 1, p = 1/2$ is equivalent to the $O(n = 1)$ model in its low temperature phase, which, as we know from the previous chapter, is critical. Hence we have identified a critical point in the θ'-model and can determine all its watermelon exponents from this equivalence. From (3.4) and (3.5) we then get $x_{\theta',L} = (L^2 - 1)/12$. This leads to the magnetic exponent $x_{\theta',1} = 0$ and the thermal exponent $x_{\theta',2} = 1/4$, from which follows $\gamma_{\theta'} = 8/7$ and $\nu_{\theta'} = 4/7$. Finally, a self contact is related to a point from which four SAWs leave. Hence we predict that $x_4 = 5/4$ describes the crossover out of criticality. This leads to $\phi_{\theta'} = (2 - x_4)\nu = 3/7$.

$\sim$

In fact, there is another way to get some of these exponents. As mentioned above, the loops representing the boundaries of the forbidden hexagons are the hulls of the triangular percolation configurations. Now because at $p = 1/2, z = 1$ the SAP has the same weight as that of any other loop and since at $p = 1/2$ the triangular percolation process is critical we arrive at the conclusion that the SAP is just equivalent to the hull of a percolation cluster at criticality. From our discussion in chapter 6 we then find again $\nu^{\theta'} = 1/D_H = 4/7$. The crossover exponent is then related to the thermal exponent ν_p of percolation. In this way we get $\phi_{\theta'} = \nu_p/\nu_{\theta'} = 3/7$. What we have used here is in fact nothing other than the relation between the critical $q = 1$ Potts model (percolation) and the low temperature $O(n = 1)$-model encountered in chapter 7. As already discussed there, the odd watermelon exponents have no immediate analogue in the percolation case.

We conclude our discussion of the θ'-model by investigating the model when $p \neq 1/2$. For $p > 1/2$, the allowed hexagons percolate and at very large length scales, p renormalises to 1. We therefore expect that in this whole regime, the SAP behaves as a non-interacting walk. When $p < 1/2$, the forbidden hexagons percolate, and therefore any SAP is confined in a finite volume. One can therefore expect it to behave like a dense walk, with $\nu = 1/2$. This is one of the reasons to believe that the collapsed phase is equal to the dense phase. We will come back to this point in section 5.

$\sim$

We conclude this section with a brief discussion of two models of interacting polymers in $d = 2$ which are closely related to the θ'-model. The first one was introduced by Seno *et al.* [142]. This model is similar to the θ'-model, but differs in the fact that now the vacancies are correlated. More specifically their correlations are that of a lattice gas, or Ising model. When one is above the Ising critical point, the vacancies only have short range correlations, and the behaviour of the SAW is therefore that of the θ'-model. At the Ising critical point, it can be shown that self avoiding rings have the properties of the hulls of Ising clusters. These are connected sets of sites in which the Ising variable is in the same state. The fractal dimension of this hull is known and is equal to 11/8. For more information about this model we refer the reader to the original reference.

Secondly, exact results are also known for the θ-model of section 8.1 on the Manhattan lattice. Bradley [143] was the first to show that on this lattice, the θ-model can also be related to percolation, but in this case the relation is to bond percolation on the square lattice. We refer the reader to the original literature for a discussion of this mapping. This map leads to the following exponents for the θ-point on the Manhattan lattice: $\nu = 4/7$, $\phi = 3/7$, $\gamma = 6/7$. Notice that the thermal exponents are the same as those on the hexagonal lattice, but that γ is different. This is essentially a consequence of the fact that on the Manhattan lattice SAWs can only become trapped at a nearest neighbour site of the starting point and therefore are like SAPs (see section 7.2). One thus understands that $\gamma = \alpha = -2\nu + 2 = 6/7$.

Later on it was realised by several authors (see for example [144]) that the interacting walk on the Manhattan lattice can be mapped onto branch 0 of Nienhuis's square lattice $O(n)$-model. This is done in the same way as for the Hamiltonian walks on the Manhattan lattice discussed in chapter 7. We first transform the SAP into an L-trail, which can then be described by putting $v = 0$ in (3.24). The square lattice model is critical along branch 0, at $u = w = 1/2$. At this point, it can be solved exactly using the Bethe *Ansatz* and one finds indeed that $\gamma = \alpha = 6/7$ in this case [145].

8.3 Numerical methods for self interacting polymers

The extension of the exact enumeration method or the transfer matrix method to the case of interacting SAWs is in principle straightforward. In the case of the exact enumeration method, one needs to determine quantities like $c_N(I)$ or a measure (such as the end-to-end distance) of the average size $R_N^2(I)$ of SAWs of N steps and I interactions. From the former quantity one can in principle determine $\mu_c(\beta)$ and the exponent γ_c. In practice it is difficult to localise the θ-point from such data. A more practical method is to determine first the average size of an N-step walk

$$R_N^2(\beta) = \frac{\sum_I c_N(I) R_N^2(I) \exp \beta I}{\sum_I c_N(I) \exp \beta I}$$

From this quantity, one can then determine a finite size estimate for $\nu_N(\beta)$. This estimate is simply given as

$$\nu_N(\beta) = \frac{1}{2} \frac{\log R_{N+1}^2(\beta) / R_N^2(\beta)}{\log (N+1)/N} \tag{8.6}$$

In a infinite system we expect that $\nu(\beta)$ has the form of a step function, but in finite systems this effect is rounded.

We now have to distinguish between two cases. In the first case (θ'-model, or θ-model on the Manhattan lattice), the location of the θ-point is known exactly. Then the analysis of the series can essentially be performed as in the non-interacting case. But more general is the situation in which the location of the tricritical point is not known exactly. Then the usual strategy is to plot $\nu_N(\beta)$ for different N values. From finite size scaling one expects that when N increases $\nu_N(\beta)$ should approach more and more closely the behaviour of a step function. A finite size estimate for T_θ^N is then given by the point where $\nu_N(\beta)$ and $\nu_{N-1}(\beta)$ intersect. These finite size estimates are then extrapolated to determine the location of the θ-point. In practice this extrapolation is usually not very simple since T_θ^N often has a rather irregular behaviour as a function of N. This makes the determination of a precise estimate of T_θ difficult. Because finite system estimates of exponent values (as in (8.6)) change very rapidly around the θ-point, the uncertainty in T_θ then leads to corresponding uncertainties in the exponent values. Exponents such as γ and ϕ are especially sensitive and thus difficult to estimate. Exact enumerations for the θ-model in $d = 2$ were performed by Privman [146] and by Ishinabe [147] for SAWs (with the respective results $\nu^\theta = 0.53 \pm 0.03$ and $\nu^\theta = 0.50 \pm 0.01$) and by Maes and Vanderzande [148] for SAPs ($\nu^\theta = 0.58 \pm 0.01$). Privman estimates $\phi_\theta = 0.64 \pm .05$, whereas Maes and Vanderzande quote $\phi_\theta = 0.9 \pm 0.1$.

Similar remarks can be made about the transfer matrix approach. When the location of the tricritical point is known, the techniques discussed in chapter 3 can be used. The most efficient method is of course a study of the spectrum of the matrix which is then analysed using conformal invariance. Duplantier and Saleur [141] studied the θ' model on strips of width up to $L = 8$ and found nice agreement with the predictions discussed in the previous section. Saleur [149] and later Veal et al. [150] studied the θ-model using the transfer matrix. In this case, since the exact location of the θ-point is not known, one has to proceed as in the case of the exact enumerations. Using, for example, phenomenological renormalisation (see (3.14)), one can obtain finite size estimates for $\nu_N(\beta)$. Using these data, one can estimate T_θ. The results obtained in this work support the values of ν^θ and γ_θ arising from the θ'-model. The value of $\phi_\theta = 0.48 \pm 0.07$ obtained by Saleur is less accurate but still consistent with $\phi_\theta = 3/7$.

$$\sim$$

Finally, we discuss how Monte Carlo methods can be modified to study interacting polymers. A simple and very effective method of the growth type was suggested by Seno and Stella [151]. A walk is grown which at each step can continue to all sites neighbouring the current position (apart from the neighbour where the walk was at the previous time step). For example, on the square lattice, there are three directions in which the walk can continue. Let us denote by n_ω the number of extra contacts that the walk acquires when it chooses to go in direction ω. We also define $n_\omega = 0$ when the walk arrives at a site which was already visited. Then a step in direction ω will be chosen with probability $p_\omega = \exp(\beta n_\omega)/(3 \exp(3\beta))$. Alternatively, the walk is stopped with a probability $1 - \sum_{\omega=1}^{3} p_\omega$. Of course, the walk also stops when a direction is chosen which was already visited before. It can be checked that the ensemble of SAWs generated in this way has the correct probability distribution. Another algorithm used by the same authors is the myopic ant algorithm where at each step one considers only those possible extensions of the walk which are allowed by self avoidance. Let us call the number of these possibilities N_c. The next step is then performed in direction ω with a probability $p_\omega = \exp(\beta n_\omega)/w$, where $w = \sum_{\omega=1}^{N_c} \exp(\beta n_\omega)$. The product of w over the different steps in a particular walk ω then gives the weight W_i of walk i, to be used in the calculation of average properties (see (4.5)). The results for $R_N^2(\beta)$ obtained in this way compare extremely well with those obtained for exact enumerations in the regime where both approaches can be applied. Of course, the Monte Carlo method allows one to investigate much longer chains. Furthermore, in this method it is also possible to calculate $Z_N^c(\beta)$. This works as follows. During the simulation we also calculate the quantity $d_N(I)$ which is the sum of all the weights W_i of walks having I contacts. An estimate for $c_N(I)$ is then given as

$$c_N(I) \approx Z_N^c(0) \frac{d_N(I)\exp(-\beta I)}{\sum_I d_N(I)\exp(-\beta I)} \qquad (8.7)$$

Multiplying by $\exp(\beta I)$ and summing over I we get

$$Z_N^c(\beta) \approx Z_N^c(0) \frac{\sum_I d_N(I)}{\sum_I d_N(I)\exp(-\beta I)} \qquad (8.8)$$

In this way from the knowledge of $Z_N^c(0) = c_N$ one can determine the partition function for general β and hence the critical exponent γ_θ. So, even though the length of the walk is fixed, we can determine $Z_N^c(\beta)$. But the crucial point is that we have to know c_N, and so what we can determine is only a ratio of numbers of walks when we are at fixed N. The data generated with this algorithm have to be carefully analysed using a combination of different techniques. We refer the interested reader to the original reference [151]. From this work, Seno and Stella estimate $\nu^\theta = 0.57 \pm 0.015, \gamma_\theta = 1.10 \pm 0.04$ and $\phi_\theta = 0.52 \pm 0.07$. These authors also locate the θ-point at $T_\theta = 1.54 \pm 0.05$. We also mention a study of the θ-point of the bond fluctuating model [152]. A Monte Carlo simulation of that model gives an exponent $\nu_\theta = 0.56 \pm 0.02$, whereas for ϕ_θ only the upper bound $\phi_\theta \leq 0.5$ is given. The interest of this work lies in the fact that it shows that at least the ν-exponent has a large universality.

In summary then, we find that from numerical work there is strong support for the idea that the exponents ν_θ and γ_θ are indeed those of the θ'-model. The situation for ϕ_θ is less convincing. We will discuss in the next section the surface critical behaviour of polymers at the θ-point. In that case, again one finds close agreement between numerical work for the θ-model and the exact predictions from the θ'-model. If one considers the bulk and surface behaviour together there is little doubt that (fortunately!) both models are indeed in the same universality class.

$$\sim$$

Monte Carlo techniques based on the generation of a Markov chain have also been extended to treat interacting polymers. The chain should asymptotically generate walks in a Gibbs state. This can be achieved by an appropriate use of the detailed balance condition (4.9). To be more concrete, let us consider again the pivot algorithm. From a given SAW W, a new configuration W' can be generated as explained in chapter 4. Suppose it is self avoiding. Then we calculate the change ΔI in the number of contacts $\Delta I = I(W') - I(W)$. When $\Delta I > 0$, W' is the next configuration in the Markov chain. When $\Delta I < 0$, W' is accepted with a weight $\pi(W, W') = \exp \Delta I$. Or, stated more elegantly, the new walk is accepted with a probability which is $\min(1, \exp \Delta I)$. With this choice, detailed balance is satisfied and we generate walks with the correct probability distribution. As the temperature is lowered, the autocorrelation time grows, because it is in general difficult to generate new chains. At very low temperatures, starting from an already compact configuration, it becomes extremely difficult to generate a new walk which is again self avoiding. From experience, it turns out that the simple acceptance/rejection method outlined above works only well down to temperatures just below the θ-point. In the compact phase, other methods have to be introduced.

In this respect, an interesting new method was recently introduced to study polymers in the collapsed phase. We will refer to it as the multiple Markov chain (MMC) method. The idea for this method was introduced by Geyer and Thompson [153], and was first used in the study of interacting polymers in $d = 3$ by Tesi *et al.* [154]. In this method one runs parallel simulations at m different temperatures $\beta_1 < \beta_2 < \ldots < \beta_m$, using e.g. the pivot algorithm for interacting polymers at a temperature β_i $(i; 1, \ldots, m)$. After every N_{sw} steps of the algorithm, one chooses two configurations at adjacent temperatures β_i and β_{i+1}, and one interchanges them. Such an interchange is called a swap. The total change in weight for the two chains is then

$$\pi_{sw} = \exp\left[(\beta_i - \beta_{i+1})(I(W') - I(W))\right]$$

A swap is then accepted with a probability which is $\min(1, \pi_{sw})$. With this procedure, one can make a very big step in phase space every now and then, and one can thus circumvent the fact that at low temperature, in a dense or collapsed phase, one gets stuck in a particular region of phase space (the problem of 'quasi-ergodicity'). Using this method, Tesi *et al.* [154] made a study of the θ-model in $d = 3$. On the cubic lattice, they found $T_\theta = 3.62 \pm 0.05$. Furthermore, they confirmed that ν_θ does indeed equal 0.5 within the numerical accuracy.

~

To conclude this section, we mention some results from the field theoretic approach to the θ-model. This was first investigated by Stephen [155] and by Duplantier [156] using a field theory for the tricritical $O(n)$-model. In that work the following expansions for the θ-exponents as a function of $\epsilon = 3 - d$ were obtained

$$\nu_\theta = \frac{1}{2} + \frac{2}{363}\epsilon^2 + \cdots$$

$$\gamma_\theta = 1 + \frac{5}{484}\epsilon^2 + \cdots \qquad (8.9)$$

$$\phi_\theta = \frac{1}{2} + \frac{3}{22}\epsilon + \cdots$$

Although originally there was some discussion about the correctness of these results, the point was settled by Duplantier [157] who found the same result using an extension of the Edwards model which also included three-body interactions. In a subsequent study, the same author [158] investigated the situation in $d = 3$ where precise predictions for the form of logarithmic corrections were obtained. As an example of the results obtained, we just mention the form of the corrections to the behaviour of the average end-to-end distance $R_N^2(T_\theta)$ and the partition sum $Z_N^c(T_\theta)$ at the θ-point

$$R_N^2(T_\theta) \approx N\left(1 - \frac{37}{363 \log N}\right)$$

$$Z_N^c(T_\theta) \sim \mu_c(T_\theta)\left(1 - \frac{49}{484\log N}\right)$$

In a recent study, Grassberger and Hegger [159] performed large scale simulations on the basis of a recursive algorithm of the growth type. Their estimate of the location of the θ-point is $T_\theta = 3.716 \pm 0.007$. This is in agreement with older estimates, but slightly outside the error bars of the estimate given above, coming from the MMC-method. More disturbing is the fact that these authors do not find agreement with the above predictions for the form of the logarithmic corrections. At this time the reason for this discrepancy is not obvious.

8.4 Adsorption and collapse

In this section we discuss a model which is a combination of the surface adsorption model of section 5.3 and the θ-model of the present chapter. We will limit ourselves mostly to the two-dimensional case. Consider therefore a model in which one has an N-step SAW (starting near a surface), which has I self contacts and V contacts with the surface (in this section V will be the number of *edges* of the SAW which are in the surface). The partition function to be studied is

$$Z_N^{c,s} = \sum_{I,V} c_N(I,V)\exp\left(\beta_1 I + \beta_2 V\right) \tag{8.10}$$

Some exact properties of this partition function have been determined by Vrbová and Whittington [160]. In figure 8.5 we draw the expected phase diagram in general d. First it is not too difficult to show that an adsorption transition must exist for all values of β_1. As long as the polymer is not adsorbed, it is clear that the location of the θ-point will not be modified, since we don't expect surface effects to influence bulk behaviour. Therefore, the phase boundary between a free bulk walk and a free collapsed walk is a straight line. When β_2 is large enough the polymer is adsorbed onto the surface. In $d = 3$, the adsorbed walk is essentially a two-dimensional object which can then collapse itself. When one compares the estimates for the location of the θ-point in $d = 2$ (square lattice) and $d = 3$ (cubic lattice) given in previous sections, one sees that the two-dimensional collapse occurs at a lower temperature. Combining

Self interacting polymers

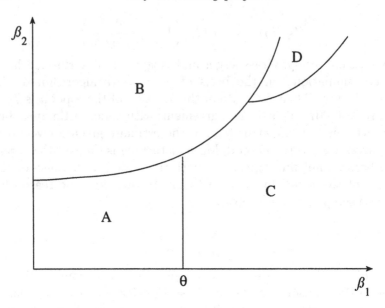

Fig. 8.5. Phase diagram of a polymer with self interactions and surface interactions. In phase A the polymer is fractal and free, while in C it is collapsed and free. In phases B and D the polymer is adsorbed. In phase D (which does not exist in $d = 2$) the adsorbed walk is collapsed.

all this reasoning one arrives at the phase diagram of figure 8.5. Sure enough, for the two-dimensional case, the phase D does not exist. The adsorption/collapse model has been extensively studied using field theoretic methods. For a discussion of these results we refer the reader to the book by Eisenriegler [85].

For the remainder of this section, we discuss the situation in $d = 2$. In fact, what we will do is discuss the θ'-model in the presence of a surface. In the original paper by Duplantier and Saleur [141] predictions were given for surface exponents at the ordinary point, which were later found to be in disagreement with numerical estimates. This led to some controversy about the universality between the θ- and θ'-models [150, 161]. The problem was solved by Vanderzande *et al.* [162], whose argument goes as follows. Consider again the θ'-model of section 8.2 and in particular consider a walk which has some steps along a surface of the

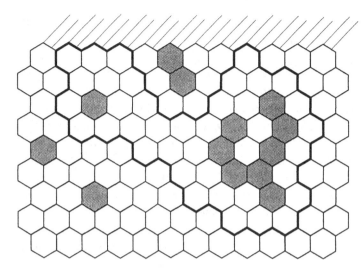

Fig. 8.6. The θ'-model near a surface. The figure shows a SAP with $V = 8, T = 4$.

hexagonal lattice (figure 8.6). After summing over the degrees of freedom of the hexagons we see that the weight of a walk is still given by p^H. The crucial modification when a surface is present lies in the expression for H. Indeed, as can be seen in the example of figure 8.6, whenever a walker makes a second step on the same hexagon situated along the surface, no new hexagon is visited. As a consequence the correct expression for H becomes

$$H = N + 1 - N_2 - 2N_3 - (V + T)/2 \qquad (8.11)$$

where T is the number of steps on bonds that are incident on a vertex at the surface. Inserting (8.11) in (8.4) we arrive at a model for a polymer which has self interactions (nearest neighbours and some next nearest neighbours) and interactions with a surface (for bonds at, and in the first layer below, the surface). More generally, we can introduce a model where walks have a weight $\exp(\beta_1(I + I') + \beta_2(V + T))$. Such a model is expected to have a phase diagram similar to the one shown in figure 8.5. The θ'-model with a surface then corresponds to one line in the phase diagram of that model, in particular the line where $\beta_2 = 2\beta_1$. Where does this line cross the line $\beta_1 = 1/T_\theta$ in the phase diagram of figure 8.5? To answer that question we look at the fraction of steps along the surface and compare the result with our knowledge of the behaviour of polymers near a surface. At the θ'-point $\beta_1 = \log 2$, a SAP becomes equivalent to the hull of a critical percolation cluster. The average fraction of steps along the surface at this point can be determined if we know the fractal dimension of the intersection between the hull and a surface. This surface fractal dimension D_s of the hull has been determined to be $2/3$ [89, 163]. This result in turn implies that the adsorbed fraction of monomers at the point $\beta_2 = 2\beta_1 = \log 4$ behaves as $m_a \sim N^{D_s/D-1}$ or $m_a \sim N^{-13/21}$. Surely this relation cannot hold in the adsorbed phase, since there $m_a \sim N$. In the ordinary regime, as mentioned in section 5.1 the surface thermal exponent y_t^s is always -1. Using the scaling form (5.23) we find for that regime that $m_a \sim N^{-\nu-1}$. Since this also doesn't correspond to the result found above, we have to conclude that the point $\beta_2 = 2\beta_1 = \log 4$ corresponds to the special point. Furthermore, comparing with (5.22) we can determine that the surface crossover exponent at the θ'-point is given by $\phi_{\theta',s} = D_s/D = 8/21$. In their paper, Vanderzande *et al.* verified these results using an exact enumeration study of the θ'-model. They furthermore determined a numerical estimate for $\gamma_{\theta',s} = 1.11 \pm 0.04$ which could be consistent with the value of $\gamma_{\theta',s} = 8/7$ allowed by conformal invariance. After this, the exact values of the watermelon exponents for the $O(n)$-model near a surface were determined, in both the special and the ordinary regime. We have already given those results in (5.25) and (5.16) respectively. Using these expressions, the above prediction for $\gamma_{\theta',s}$ can be confirmed. In the ordinary regime one finds from (5.16) $\gamma_{\theta',s} = 4/7$. This value is indeed consistent with numerical estimates for the θ-model [150, 164]. Finally we mention the work of Foster *et al.* [165] who performed an exact enumeration study of the surface behaviour of the θ-model and verified that in this case the exponents predicted from the θ'-model are also consistent with the numerical data. This then leaves little doubt that the θ- and θ'-models are indeed in the same universality class.

$$\sim$$

By this point we have encountered many exponents for the two-dimensional SAW. In table 8.1 we give an overview of what are now considered to be the exact exponents of that model in various situations.

Table 8.1. *Overview of exponents for the SAW in $d = 2$*

	ν	γ	ϕ	γ_s(ordinary)	γ_s(special)	ϕ_s
good solvent	3/4	43/32		61/64	93/64	1/2
θ-point	4/7	8/7	3/7	4/7	8/7	8/21

8.5 The collapsed phase

We now turn to a study of the interacting SAW model at low temperatures, i.e. below the θ-point. In that region the walk is collapsed and thus we should have $\nu = 1/d$. The study of this regime has been boosted in recent years by a suggestion by Owczarek *et al.* [166] that in this regime the partition function $Z_N^c(\beta)$ should have the following form

$$Z_N^c(\beta) \sim \mu_c(\beta)^N \mu_1(\beta)^{N^\sigma} N^{\gamma-1} \quad 1/\beta < T_\theta \quad (8.12)$$

already familiar from our study of dense polymers. Besides an eventual correspondance between collapsed and dense polymers there is a simple reason to expect a form like (8.12). At low temperatures the leading contribution to $Z_N^c(\beta)$ comes from those walks (figure 8.7) which have as many self contacts as possible. In the interior of such a walk one can indeed achieve one contact for each bond, but near the surface a number of such contacts necessarily have to be missed. We can then estimate that for such a collapsed walk $I = I_{\max} = N - cN^\sigma$ where $1/\sigma - 1$ corresponds to the fractal dimension of the surface. For a sphere-like configuration then $\sigma = 1/d$. If the number of walks with this maximal number of self interactions is c_N^m we find that at low temperature

$$Z_N^c(\beta) \sim c_N^m \exp(\beta N) \exp(-\beta c N^\sigma) \quad (8.13)$$

which is indeed of the form (8.12) irrespective of whether the walks with $I = I_{\max}$ are those of the dense phase or not.

The interesting question is whether the collapsed walks correspond to dense walks and, at $T = 0$, eventually become Hamiltonian. A study of the θ-model at low temperatures using a series expansion method was performed by Bennett-Wood *et al.* [167]. They determined the ratio of SAWs to SAPs at sufficiently low temperatures. If these walks are described by the low temperature

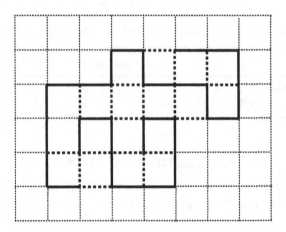

Fig. 8.7. A walk with many self contacts.

phase of the $O(n)$-model one expects to find numerical evidence for the prediction (7.9). In fact the value for the exponent which was found was lower than 19/16. The estimate is 0.92 ± 0.09. It is not clear how to interpret this result. We know from the discussion in the previous chapter that the γ-exponent in dense phases may be highly non-universal. For example, if the interacting walk collapses into a Hamiltonian walk one expects a γ-exponent of 1.0444 ± 0.0001 (on the square lattice), which may just be consistent with the result of Bennett-Wood *et al.*. In judging these results one also has to take into account that it is difficult to perform reliable numerical work in the low temperature phase. Series don't get very far and the Monte Carlo method has ergodicity problems. In recent work, Vanderzande studied the collapsed walk on the Manhattan lattice, where because walks cannot get trapped, it is possible to create rather long walks with a growth algorithm of the type discussed in section 8.3. If one assumes that the walk collapses into a Manhattan Hamiltonian walk, then combining (8.13) and (7.13) we predict

$$Z_N^c(\beta) \sim \exp\left[(\beta + G/\pi)N\right] \left[(1 + \sqrt{2})^{-1}\exp{-\beta c}\right]^{\sqrt{N}} \dots$$

The numerical work gave very good agreement with this form, although unfortunately no reliable estimate of the γ-exponent could be determined. Thus, the little evidence available now seems to suggest that the low temperature properties of interacting walks are those of Hamiltonian walks on the corresponding lattice. Further work remains to be done to settle this problem.

8.6 A model with direction dependent interactions

As our next example of a self interacting walk, we discuss a model which has recently been introduced [168]. In this model the SAW is given a direction and the interactions between nearest neighbour *edges* depend on whether the arrows on these bonds are parallel or anti-parallel (figure 8.8). This kind of model, but without the self-interaction, has been introduced to describe oriented polymers, such as $A - B$ polyester [169]. Consider a model in which we count the number of parallel contacts I_p and the number of antiparallel contacts I_a and give them different weights when calculating the partition function $Z_N^d(\beta_a, \beta_p)$

$$Z_N^d(\beta_a, \beta_p) = \sum_{I_a, I_p} c_N(I_a, I_p) \exp(\beta_a I_a + \beta_p I_p) \qquad (8.14)$$

The recent interest in this model stems from the fact that for $\beta_a = 0$ the model is predicted to have a γ-exponent which varies continuously! This follows from work by Cardy using field theoretic methods and conformal invariance [168]. A lot of work has been devoted to this model in order to determine its phase diagram and to find evidence for the continuously varying exponents. In this section we review that work.

From the partition function (8.14) we can construct a free energy in the usual way

$$f^d(\beta_a, \beta_p) = \lim_{n \to \infty} \frac{1}{N} \log Z_N^d(\beta_a, \beta_p) \qquad (8.15)$$

To be precise, there is no general proof that this free energy exists, but there is also no physical reason why this shouldn't be the case. We now show that as a function of β_p, the model must have a transition (we follow the arguments in [170]). Consider first the case $\beta_a = 0$. For all $\beta_p \leq 0$ we have the inequality $Z_N^d(0, \beta_p) \leq Z_N^d(0, 0) = c_N$. On the other hand the subset of SAPs cannot have any parallel contacts and therefore $Z_N^d(0, \beta_p) \geq c_N(I_a, 0) \geq q_N$. Since the connective constant for SAWs and SAPs is the same, this argument shows that $f^d(0, \beta_p) = \log \mu_2, \ \forall \beta_p \leq 0$. Then consider

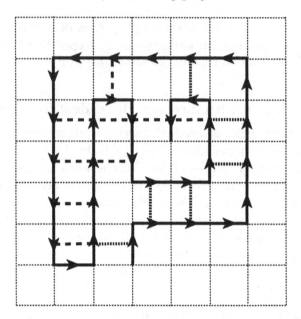

Fig. 8.8. A SAW with direction dependent interactions.

the case that $\beta_p \rightarrow \infty$. The partition sum Z_N^d will be dominated by spiral-like walks which have as many parallel contacts as possible. In such a spiralling walk $I_p \sim N$, up to surface corrections. Therefore, a lower bound for $f^d(0, \beta_p)$ is given by $\beta_p + \log \mu_{sp}$ where μ_{sp} is a connective constant for spiralling walks. As in the case of the polymer adsorption, these arguments then show that $f^d(0, \beta_p)$ must be non-analytic at some $\beta_p \geq 0$. Similar reasonings can be made for the case $\beta_a \neq 0$. One can then show that the connective constant at fixed β_a is at least constant up to $\beta_p = \beta_a$, and that there must be a phase transition at some $\beta_{p,c}(\beta_a) \geq \beta_a$. Finally, take into account the fact that along the line $\beta_a = \beta_p$ the model with direction dependent interactions coincides with the usual θ-model which has a phase transition at the θ-point. This leads us to the phase diagram of figure 8.9.

In recent years the model (8.14) has been studied by exact enumerations [170], Monte Carlo methods [171] and transfer matrix techniques [172, 173], especially along the line $\beta_a = 0$. So far little evidence of continuously varying exponents has been found. From

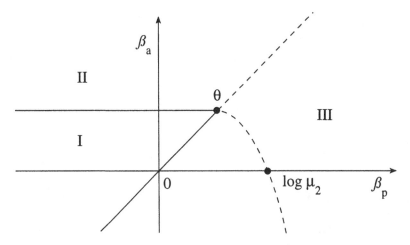

Fig. 8.9. Phase diagram of the SAW with direction dependent interactions. Phase I is the non-interacting SAW phase, while in the other phases polymer is collapsed. The full line indicates a line of θ-transitions in the universality class of the Manhattan lattice θ-point. The dashed lines correspond to first order transitions. The exact form of the phase diagram in the neighbourhood of the θ-point is unknown. What is shown here is the simplest scenario, consistent with exact and numerical work. The intersection of the first order line with the axis $\beta_a = 0$ is located at $\beta_p = \log \mu_2$ (see [170]).

the Monte Carlo work, it has been found that $\beta_{p,c}(0) \approx \log \mu_2$. At that point the walk collapses through a first order transition into a compact spiral phase. Furthermore, from extensive transfer matrix studies Trovato and Seno [173] find that along the line $\beta_p = -\infty$ the collapse at $\beta_a = 1/T_\theta$ is in the universality class of the θ-point on the Manhattan lattice, i.e. with $\gamma_\theta = 6/7$. This makes sense since the Manhattan walks form a subset of the walks without parallel interactions. The same behaviour is found all along the line $\beta_a = 1/T_\theta$, up to the ordinary θ-point. This is indeed what is to be expected from the above arguments on the behaviour of f^d. But since at the θ-point we are in another universality class, at $\beta_a = 1/T_\theta$ the free energy certainly becomes non-analytic at the point $\beta_a = \beta_p$. So the simplest scenario is that shown in figure 8.9 in which the usual θ-point is a higher order critical point where two lines of first order transition and one line of tricritical points meet. Further work has to be done to verify these conjectures, and to understand why the continuously varying critical exponent has not been found numerically.

8.7 Polymers with long range interactions

Imagine a long polymer chain where each of the monomers carries a charge q. These charges interact through their Coulomb potential. What will be the influence on the critical behaviour of the polymer? This and related questions will be shortly discussed in this section.

Charged polymers with long range interactions are usually referred to as *polyelectrolytes*. In recent years quite some interest has been given to a related kind of polymer called a *polyampholyte*. In such a polymer the charges may have different signs so that both repulsive and attractive interactions are present. The study of these polymers in turn is a simple case of the study of proteins, which are important polymers occurring in, for example, living systems.

For the moment we go back to the polyelectrolytes. In general when such a polymer is immersed in a solvent, counterions of the solvent will tend to screen the monomer charges. This is similar to the Debye screening in solids. Therefore the true long range interaction will effectively be cut off above some screening length. What we will say in the following discussion will only hold when the size of the polymer is below this screening length so that the interactions are effectively long range. We leave it to the physical chemists to determine to which real world situations our somewhat academic discussion applies.

To determine the behaviour of polyelectrolytes we follow an argument of RG type which was first given by de Gennes and his coworkers [174]. The argument is of interest because it can also be applied to polyampholytes. For the moment we will neglect the excluded volume interactions since we will show at the end of our discussion that they don't matter anyhow. The Coulomb interaction energy in general dimension d is of the form

$$V_C = c(d)\frac{q^2}{\epsilon r^{d-2}} \tag{8.16}$$

where $c(d)$ is some d-dependent constant, and ϵ is the permittivity of the solvent. Distances r between charged monomers will be measured in some length unit l. The dimensionless constant measuring the strength of the Coulomb interaction is therefore

$$g_C = \frac{c(d)q^2}{\epsilon l^{d-2}k_B T} \qquad (8.17)$$

As usual we are interested in the exponent ν which relates distance to the number of monomers. Suppose we are somehow able to perform a RG calculation in which sizes are rescaled by a factor b. The number of monomers will then be rescaled by a factor $b^{1/\nu}$. The coupling constant g_C will under this rescaling transform as

$$g'_C = g_C b^{2/\nu} b^{2-d} \qquad (8.18)$$

where the first term comes from the rescaling of the total charge (which just gets multiplied by b) and the second from the rescaling of l. When g_C is very small we expect that the behaviour of the polymer will be that of a random walk; therefore in that regime $g'_C/g_C = b^{6-d}$. In the very strong coupling regime, on the other hand, the Coulomb repulsion will stretch the polymer completely and as a consequence $\nu = 1$. Therefore in that regime $g'_C/g_C = b^{4-d}$. As a consequence, when $d < 4$, g'_C will always be bigger than g_C and under renormalisation we end up in the strong coupling regime where $\nu = 1$. Similarly, for $d > 6$ we are always driven to the very weak coupling limit, and the behaviour of the polymer is always ideal. Finally, for $4 < d < 6$, there will be a fixed point at some value $g^\star_C$. We immediately find from (8.18) that in that range of spatial dimensions ν is given by

$$\nu = \frac{2}{d-2}$$

These predictions for ν should not be modified by self avoidance, since for $d < 4$ the polymer is already fully stretched and for $d > 4$ self avoidance is known to be irrelevant. Although de Gennes' argument is very heuristic, it has been shown that it predicts the correct value of the ν-exponent [174].

~

We now turn to the polyampholyte case. Simple models for these have been studied by several authors in recent years. The most extensive work has been done by Kantor and Kardar [175,

176] and here we give a brief overview of their results. As mentioned above, in polyampholytes charges of different signs occur. As a simple model we will assume that these charges are randomly distributed along the polymer and can take on the two values $\pm q$ with equal probability. These charges are taken as quenched random variables, i.e. they are fixed for a given polymer. On average, the charge of a polymer will therefore be zero but a typical polymer will be charged. This charge will be of order $q\sqrt{N}$. It has been found that the behaviour of the polymer will strongly depend on the value of the total charge Q_t.

We can now go through de Gennes' argument for polyelectrolytes and hope that in this case too, we get the exact result. Because of the randomness the RG recursion (8.18) is now replaced by

$$g'_C = g_C b^{1/\nu} b^{2-d} \qquad \text{random case} \qquad (8.19)$$

which implies that the electrostatic interaction, just as the excluded volume one, becomes irrelevant in $d > 4$. When $d < 4$, we predict that

$$\nu = \frac{1}{d-2} \qquad (8.20)$$

implying that in $d = 3$ the polymer is stretched. Therefore, in dimensions $d \leq 3$ the polymer is always stretched. This simple picture is partly confirmed by numerical simulations of the model. Two main techniques have been used in studying these polymers. The first one is a Monte Carlo simulation using the fluctuating bond method [175]. Such a technique can be expected to work correctly for high temperatures or weak couplings. At low temperatures, the free energy of a polyampholyte is dominated by one or at most a few ground states. With a Monte Carlo method it is highly unlikely that one will encounter these states and therefore the free energy and average properties are poorly determined. This is a general problem for random systems. At low temperature, one therefore often uses exact enumeration techniques. But here one is limited by the total length N which one can study on, for example, a hypercubic lattice (see chapter 2). Furthermore, one has to perform an average over all realisations of randomness. This limits the sizes which can be reached even further; in their work Kantor and Kardar were able to reach up to $N_{\text{max}} = 13$ [176].

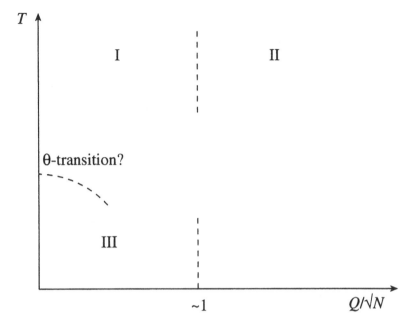

Fig. 8.10. Qualitative phase diagram of polyampholytes as a function of temperature and total charge (adapted from reference [177]).

Besides these numerical techniques one can try to apply approximate theories such as the Debye–Hückel theory for electrolytes [177]. It is thus clear that the current understanding of the behaviour of polyampholytes is still far from complete. A Monte Carlo technique which is able to probe systems at low temperatures would be most welcome; maybe the multiple Markov chain method discussed in section 8.3 could be such a method.

The phase diagram which results from combining the results of all the approximate methods discussed so far is shown in figure 8.10. As already mentioned, the phase diagram depends in a crucial way on the total charge. At high temperatures, perturbative methods show that when $Q_t < q\sqrt{N}$ the polymer contracts whereas in the reverse case the polymer stretches and its behaviour is consistent with the predictions (8.20) of the Flory-type theory. Figure 8.11 shows the results of numerical simulations at high temperatures which show this stretching effect very clearly. There

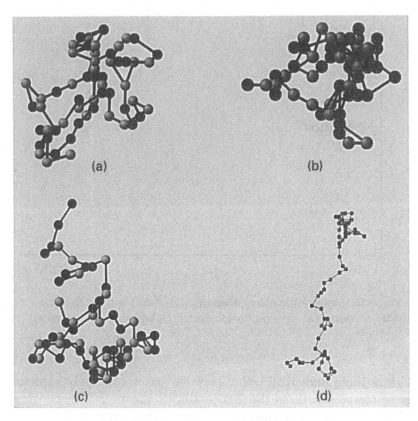

Fig. 8.11. Spatial conformations of polyampholytes of 64 monomers for values of $Q_t =$ (a) 0, (b) 4, (c) 8 and (d) 16. Dark and bright shades indicate opposite charges (from reference [176]).

exist arguments that this stretching transition occurs all the way down to zero temperature [175]. For the weakly charged walk it is possible that there is a θ-like transition [176]. Indeed, in the case of polyampholyte models with short range interactions, there is strong evidence that for $Q_t \approx 0$, these polymers behave just as homopolymers which have a θ-transition from an open fractal to a compact phase [178]. In the case of long range interactions, there may be a similar transition. Some evidence for this has been seen in exact enumeration work. We know of no numerical estimates for exponents in these condensed phases. In fact, there exist a prediction from Debye–Hückel theory [177] that the polyampholyte is more open at low temperatures than at high temperatures (always speaking of the case that the total charge is less than $q\sqrt{N}$). Clearly, a lot of work remains to be done before these interesting polymers are understood.

~

The problems one encounters in studying polyampholytes are simple compared with those that show up in studying proteins. Proteins [179] are heteropolymers that are built from 20 different types of monomers (the naturally occurring amino acids). They are weakly charged but a first difference from polyampholytes is the fact that the proteins are definitely not random sequences of amino acids. Furthermore many different interactions exists between the amino acids constituting the protein.

Proteins always seem to exist in one configuration which is rather compact and which is usually referred to as the native state. In this configuration, the protein has a definite shape. This shape is crucial for the correct working of the protein in its biological environment. The shape varies from protein to protein and depends on the exact sequence of amino acids constituting the protein. This is explained as follows: the amino acid sequence (the so called primary structure) determines the strength of the different interactions in the protein, and thus the total energy of the polymer depends in a complicated way on its three-dimensional structure. Probably, the native state is the configuration in which the energy is minimised. The 'protein folding' problem, then, consists essentially of two problems. First, can one, given the amino acid sequence, determine the structure (the so called tertiary structure) of the native state or as a reverse problem, can one determine from the native state the amino acid sequence which folds into that native state? Solving this reverse problem would be of great use in pharmacology since it could lead to the design of proteins which are suitable for particular medical applications. The second part of the protein folding problem is a dynamical one. In biological cells, proteins are built as linear chains by subsequent addition of amino acids. How does this linear structure fold into the native state, and how can the protein so 'quickly' find its lowest energy state? In this respect, proteins seem 'cleverer' than humans who have only very inefficient algorithms to solve this complex problem. Understanding this part of the protein folding problem would be of great importance in the physics and mathematics of many random systems. It seems clear that a complete understanding of the behaviour of proteins cannot be achieved through the simple models we have used so far in this book. Such models can at most lead to the understanding of some general principles involved in protein folding.

The simplest model of a protein is undoubtedly the HP model introduced by Dill and coworkers [180]. In this model, which only aims to describe the equilibrium properties of proteins, a protein is considered as a SAW consisting of two type of monomers denoted as H (hydrophobic) and P (polar). The hydrophobic monomers tend to avoid contact with the solvent and therefore have the attractive interaction present in the θ-model. For P–P pairs or P–A pairs there is no interaction. The interactions are only short ranged. A lot of numerical work has been devoted in recent years to study this model, again using exact enumeration and Monte Carlo techniques. Interesting questions (besides the usual ones about entropy and exponents which are not so relevant for proteins) to ask about this model are: for a given sequence of H and P monomers, what is the degeneracy of the ground state? For a given ground state (i.e. a particular configuration of a SAW) how many H–P sequences lead to that ground state?

In general, the results found are in surprisingly good agreement with what one knows about real proteins. For example, for many sequences of H and P monomers, the number of ground states is usually one (apart from obvious rotations) or at least very small. The ground states are rather compact with the H monomers on the inside.

Although these studies are of very great interest, at this moment they are at the limit of what the methods described in this book allow us to understand. Here lies the real frontier for the lattice models of polymers.

9
Branched polymers

So far we have studied polymers with a linear structure. This is a consequence of the fact that the monomers have a functionality of two, which means that each monomer can bind to two other monomers. Branched polymers occur when the functionality of the monomers is higher. Branched polymers (BP) can have a fixed topology, meaning that they consist of a fixed number of branches and nodes. We will see that the properties of such polymers are still closely linked to those of linear polymers (section 9.1). In some cases it is more appropriate to consider the functionality as random and to describe the polymers as lattice animals. These lattice animals can again be described by the Potts model, but a description using field theory will turn out to be more instructive. We will consider first a branched polymer in a good solvent and later the phenomena of adsorption and collapse for these polymers. Branched polymers also turn up in the study of vesicles which are simple models for cell membranes.

9.1 Branched polymers of fixed topology

As a first simple model for a branched polymer we can consider a star polymer, which consists of N_a arms which each have the same number N of monomers. These arms are modelled by SAWs (figure 9.1). In the same figure we also show a more arbitrary polymer which can be described as a graph with specified vertices and edges (more precisely, one can say that the polymer is an embedding of the graph). We will denote in general by N_a the number of branches of the polymer. Let the graph have V_i vertices of degree i (the degree of a vertex is the number of edges incident on that vertex). The total number of vertices in the graph is $V = \sum_i V_i$. We will consider two related situations. First, we will consider the

149

Branched polymers

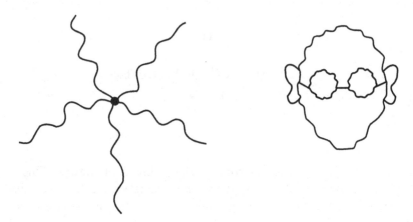

Fig. 9.1. A star and a more complex branched polymer.

case in which the number of monomers N in each branch is fixed.
Each arm is described by a SAW, and all the arms have to be
mutually avoiding. This will be referred to as the homogeneous or
monodisperse case. In the second situation, only the total number of monomers is fixed (the polydisperse situation). We ask the
familiar questions; what is the number of configurations for the
branched polymer, how does its average size depend on N and so
forth?

We begin by outlining a proof given by Soteros *et al.* [181] which
shows that for the non-uniform case the connective constant is the
same as that of SAWs. First, note that from any branched polymer
of m monomers we can construct another branched polymer of
N monomers with the *same* topology by concatenating it with a
polygon of $N - m$ monomers (see figure 9.2). The number of BP
configurations can thus be bounded from below by the number of
polygons of $N - m$ monomers. On the other hand each branched
polymer can be decomposed into N_a SAWs of length $N_1, \ldots, N_{N_a}$.
An upper bound to the number of BPs is then given in terms of
the number of SAWs of total length $N = N_1 + \ldots + N_{N_a}$. Since
SAPs and SAWs have the same connective constant one finds from
this upper and lower bound that the number g_N^f of BPs of fixed
topology (and total number of monomers N) also has the same

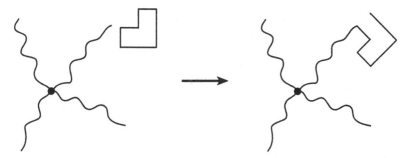

Fig. 9.2. Concatenation of a branched polymer and a polygon (schematically).

connective constant. We thus arrive at the conclusion that

$$g_N^{\mathrm{f}} \sim \mu^N N^{\gamma_g - 1} \tag{9.1}$$

where γ_g depends on the specific form of the BP. Later, Soteros [182] also gave a proof that for the monodisperse case the connective constant is that of the SAW. The exponent γ_g will depend on whether we are in the monodisperse or the polydisperse situation, but both can be easily related (see below). We will denote the two different γ-exponents by superscripts m and p respectively.

A similar result for the connective constant together with an expression for the γ-exponent was derived by Duplantier [183], who showed how to relate γ to the watermelon exponents of the $O(n)$-model. These are known in $d = 2$ and one can derive expressions for them in higher dimensions using the ϵ-expansion. We first introduce the notation $O_L(r)$ for the 'operator' which has critical exponent x_L [23]. To be more concrete $O_1(r)$ corresponds to the spin $\vec{s}_r$, $O_2(r)$ corresponds to the energy in the bond at r and so on. We call L the degree of the operator $O_L(r)$. Next, we calculate the multiple correlation function $\langle \prod_\alpha O_{L_\alpha}(r_\alpha) \rangle$ in the $O(n)$-model. The product is over all vertices of the graph representing the BP and L_α is the degree of the operator at r_α. Endpoints have to be included as vertices of degree 1. This correlation function is given by the sum of all BPs going through the points r_α. The number of self avoiding lines going out of the point r_α is L_α. If there are N_{a} branches in the polymer with respective lengths $N_1, \ldots, N_{N_{\mathrm{a}}}$

(with total length $N_t = \sum_i N_i$), we have

$$\langle \prod_\alpha O_{L_\alpha}(r_\alpha) \rangle = \sum_{N_1} \cdots \sum_{N_{N_a}} v^{N_t} g^{f,p}(\{N_k\}, r_\alpha) \qquad (9.2)$$

where in an obvious notation $g^{f,p}(\{N_k\}, r_\alpha)$ gives the number of BP of fixed topology going through the set of points r_α, in which branch k has N_k monomers. Next we integrate over all but one of the positions r_α. This integration modifies the last term on the right hand side of (9.2) into $g^{f,p}$. On the other hand, we expect that near criticality, the correlation function in (9.2) will decay as some typical distance D to the power $-\sum_i V_i x_{L_i}$, multiplied by a scaling function of D/N^ν. Combining all these arguments, and doing an inverse Laplace transform, we find that $g^{f,p}$ is indeed of the form (9.1) with

$$\gamma_G^p = \nu \left(2(V-1) - \sum_i V_i x_{L_i} \right) \qquad (9.3)$$

As an example, for the simplest case of a linear polymer, we have $V = 2$, and $V_1 = 2$. In this way, we recover the well known formula for $\gamma = \nu(2 - 2x_1)$.

As a next step, we want the exponent γ_G^m for the monodisperse case [183]. It is defined in

$$g_N^{f,m} \sim \mu^{N_a N} N^{\gamma_G^m - 1} \qquad (9.4)$$

We can estimate the ratio $g_N^{f,p}/g_N^{f,m}$ as a geometrical factor, i.e. as the number of possible ways to have N_a arms of respective length $N_1, \ldots, N_{N_a}$ but with a total length of N:

$$\int \delta \left(N - \sum_{i=1}^{N_a} N_i \right) dN_1 \ldots dN_{N_a}$$

This integral is proportional to $N^{N_a - 1}$. This leads to the conjecture that

$$\gamma_G^m = \gamma_G^p + 1 - N_a \qquad (9.5)$$

Combining (9.3) and (9.5) then gives an expression for the γ-exponent of an arbitrary BP of the homogeneous type. As an example, a star of N_a branches of equal length has an exponent

$$\gamma_{N_a}^m = \nu \left(2N_a - N_a x_1 - x_{N_a} \right) - N_a + 1 \qquad (9.6)$$

Using (3.5) (for $n \to 0$) then finally gives

$$\gamma_{N_a}^m = \frac{68 + 9N_a(3 - N_a)}{64} \tag{9.7}$$

These predictions are in very good agreement with numerical results from an exact enumeration calculation [184]. These give e.g. $\gamma_3 = 1.07 \pm 0.02$ and $\gamma_4 = 0.52 \pm 0.04$. For the non-uniform case, we get $\gamma_3 = 49/16$ and $\gamma_4 = 7/2$. Although simulations by Zhao and Lookman [185] were not completely consistent with these predictions, more recent studies of larger stars by Grassberger [186] give results in very close agreement with (9.6).

We mention that Duplantier [183] also derived similar results to first order in the ϵ-expansion. Finally, note that since the critical behaviour of these 'simple' branched polymers is that of the SAW, the ν-exponent also is that of the $O(n = 0)$-model.

9.2 The critical behaviour of lattice animals

In this section we study random BPs where the topology is not fixed. These BPs are modelled by lattice animals, which we encountered earlier in our study of percolation. The description of BPs using lattice animals was pioneered by Lubensky and Isaacson [187]. There are several different kind of lattice animals; we can either count them by their number of bonds or edges s (bond animals), or by their number of sites v (site animals). For a SAW $v = s + 1$. We will denote by g_s and g_v, respectively, the number of lattice animals of s bonds or v sites. In this section we will mainly consider the bond case. One can also study a subset of lattice animals which are 'trees', i.e. animals containing no loops. From Euler's relation (6.3) it follows that for trees $v = s + 1$, so there is no need to distinguish between trees counted by vertices and by bonds. In this respect trees are more similar to SAWs. We will denote the number of trees by t_v. Finally we can distinguish between animals for which each bond between two sites has to be part of the animal ('strong embedding') or a number of these bonds are absent ('weak embedding'). We will use capital letters for the strong embedding case: G_s and G_v. Such a distinction is of course not necessary for trees. The different classes of lattice animals are shown in figure 9.3.

Branched polymers

$v=11, s=12$ $v=11, s=15$ $v=11, s=10$

Fig. 9.3. Different classes of lattice animals. From left to right we
show: a weakly embedded animal, a strongly embedded one, a tree.

We have encountered the quantity g_s already in our study of
percolation. Its asymptotic behaviour is believed to be of the form
(6.8)

$$g_s \sim \mu_a^s s^{-\theta} \tag{9.8}$$

This is the form familiar from linear polymers. That the leading
term in the expression for g_s is exponential can again be argued on
the basis of concatenation arguments [188]. Expressions similar to
g_s are thought to be valid also for g_v, G_s, G_v and t_v. For historical
reasons the exponent is denoted θ in the case of animals. Since
for animals there is no distinction between closed or open as for
SAWs, one cannot say whether θ resembles γ or α. We expect θ
to be a universal exponent which doesn't depend on whether the
animal is a tree or on whether it is counted by vertices or bonds.
We will however encounter some subtle effects in the behaviour of
θ at the end of this section. Besides θ we are also interested in the
exponent ν which, as usual, describes how a measure of the size
of the animal (e.g. its radius of gyration R_s) diverges for large s.

The most extensive studies of lattice animals have been performed using exact enumerations both in $d = 2$ and $d = 3$. In
table 9.1 we give a summary of the connective constants which
have been obtained for all the different models of branched polymers on the square and on the cubic lattice (these data are taken
from [189]–[192]).

What about the exponents θ and ν? The situation for these
exponents is opposite to that of SAWs in the sense that 'exact'
conjectures for these exponents exist in $d = 3$. Also, in $d = 2$, θ is
known. Unknown is the exponent ν in $d = 2$. These remarkable re-

Table 9.1. *Connective constants for different lattice animal models*

type of animal	$d = 2$	$d = 3$
site (strong)	4.063	8.339
site (weak)	5.496	11.404
bond (strong)	3.877	7.907
bond (weak)	5.212	10.633
trees	5.140	10.496

sults come from a relation between lattice animals in d dimensions and the exponents of the so called Lee–Yang edge singularity in $d - 2$. This relation was first demonstrated by Parisi and Sourlas [193]. It is the analogue of de Gennes' theorem for SAWs. As we will see below, however, this result does not have the same status as that for the linear polymer. Nevertheless, there is by now little doubt that the Parisi–Sourlas exponents are the exact ones.

Let us first discuss the Lee–Yang edge singularity [194]. Consider therefore an Ising model at an inverse temperature K and in a magnetic field H. We will be interested in the zeros of the partition function $Z(K, H)$ when H is considered to be a complex variable $H = H_1 + iH_2$. As a concrete example, let us take an Ising model in $d = 0$, i.e. just a single spin in a field. The partition function is $Z_{d=0}(H) = 2 \cosh(H_1 + iH_2)$ which can be rewritten as

$$Z_{d=0}(H) = 2 \left(\cosh H_1 \cos H_2 + i \sinh H_1 \sin H_2 \right) \quad (9.9)$$

The zeros of the partition function are all purely imaginary $H = i(2n + 1)\pi/2$ for any integer n. We will denote by H_0 the zero closest to the real axis, i.e. $\pm H_0 = \pm i\pi/2$. We next consider the magnetisation $M(H) = \tanh H$ for H close to $+H_0$. We find

$$M = \frac{\tanh H_1 + i \tan H_2}{1 + i \tanh H_1 \tan H_2}$$

$$= \frac{1 + iH_1(H_2 - \pi/2) + \cdots}{H_1 + i(H_2 - \pi/2)}$$

If we define $h = H - H_0$, we see that $M = 1/h + \ldots$. In general, one defines an exponent $\sigma(d)$, which relates M and h as $M \sim h^{\sigma(d)}$. We see that $\sigma(0) = -1$. In $d = 1$, the calculation of σ is again

Table 9.2. *Critical exponents for BPs*

d	2	3	4	8
θ	1	3/2	11/6	5/2
ν	?	1/2	5/12	1/4

straightforward. The partition function of a one-dimensional Ising model of N spins is [195]

$$Z_{d=1}(K, H) = \exp{(KN)} \left[\cosh H \pm \sqrt{\sinh^2 H + \exp{-4K}} \right]^N \quad (9.10)$$

The magnetisation per spin becomes in the thermodynamic limit

$$M = \frac{\exp{(K)} \sinh H}{\sqrt{\exp{(2K)} \sinh^2 H + \exp{(-2K)}}} \quad (9.11)$$

The denominator can only become zero when H is purely imaginary. In particular, one obtains $H_0 = i \arcsin[\exp{(-2K)}]$. Expanding M for small values of $h = H - H_0$, one finds that $M \sim 1/\sqrt{h}$, which leads to $\sigma(d = 1) = -1/2$.

One can go on to calculate σ in higher dimensions. Using conformal invariance Cardy [196] obtained $\sigma(d = 2) = -1/6$. He furthermore identified the central charge of this critical point to be $c = -22/5$. Fisher [197] was able to show that the upper critical dimension for the Lee–Yang problem is $d = 6$, above which one has $\sigma = 1/2$. He also derived a field theory that describes the Lee and Yang edge singularity near this upper critical dimension. This allows the calculation of σ in an ϵ-expansion around $d = 6$.

We can now go back to BP. The result of Parisi and Sourlas relates the exponents θ and ν for the BP in d dimensions to the exponent σ of the Lee–Yang problem in $d - 2$. Specifically, their results are

$$\theta(d) = \sigma(d - 2) + 2 \quad (9.12)$$

$$\nu(d) = \frac{\sigma(d - 2) + 1}{d - 2} \quad (9.13)$$

This leads to the predicted exponents for BPs shown in table 9.2. As we see the main gap is $\nu(d = 2)$ where (9.13) leads to an indefiniteness.

There exists no simple way to derive the Parisi–Sourlas result (see also the discussion in [23]). The starting point is a continuum theory of an $O(n)$-like model, which for $n \to 0$ gives diagrams which are lattice animals. Parisi and Sourlas then study this field theory using the RG near its upper critical dimension. Within this approximation, where one can omit all but the most singular diagrams, the remaining diagrams turn out to be the same as those for a magnetic system in an imaginary and random magnetic field when this model is studied with the replica trick (which we encountered in chapter 1). Finally, this model can be related to Fisher's field theory for the Lee–Yang singularity. The relations (9.12) and (9.13) are thus only known to be true near the upper critical dimension. Yet, as we will discuss below, the predictions are in very good agreement with numerical estimates. The current belief is therefore that the Parisi–Sourlas equations are exact.

~

How can one fill in the gap in table 9.2 ? From the study of the previous chapters one may be inclined to think that the ν-exponent of BPs is known in $d = 2$ from the Coulomb gas or from conformal invariance. This is not the case. As we will see in section 9.4, BPs can be described by an extension of the Potts model, but the relation between that model and the Coulomb gas is not completely known. Furthermore, numerical evidence indicates that branched polymers are not conformally invariant [198]. At the end of section 5.2 we discussed how conformal invariance allows the calculation of the value of the surface exponents for a polymer in a wedge with an opening angle α. Exact enumerations indicate that branched polymers do not follow that relation. Also, the relation (3.17) does not seem to hold for branched polymers. It is not very clear why this is the case. Because of this, no watermelon exponents can be given for lattice animals or trees.

In order to determine the exponent ν in $d = 2$ and to verify the predictions of the theory of Parisi and Sourlas we therefore have to rely on numerical work.

Exact enumeration work gives values for ν around 0.64 [199], whereas in $d = 3$ the values for ν are more scattered but consistent with $\nu = 1/2$ [189]. Exact enumerations are one of the few methods that can be used to calculate the exponent θ. The available results confirm the values coming from the Parisi–Sourlas theory.

The most precise determinations of ν in $d = 2$ have been obtained by the transfer matrix calculations of Derrida and de Seze [200], and Derrida and Stauffer [201]. The transfer matrix method needs no essential modification when applied to branched polymers. From their work, in which strips with widths up to $W = 10$ were considered, Derrida and Stauffer obtained the very precise estimate $\nu = 0.64075 \pm 0.00015$. Furthermore they obtained an independent estimate of the connective constant for site animals in the strong embedding case. Their result is $\mu_a = 4.06256 \pm 0.00017$ which can be compared with the series result in table 9.1. The relation (3.17) between the gap in the spectrum of the transfer matrix and the x-exponent doesn't seem to hold for branched polymers [201], another indication that conformal invariance doesn't hold, so no estimate of θ can be obtained from that method.

Finally, we look at Monte Carlo methods for branched polymers. At this moment, there exist far fewer methods for branched than for linear polymers. Furthermore, almost all methods work only for trees, which are simpler than general lattice animals, and more similar to SAWs. A simple method which works at fixed fugacity was proposed by Duarte [202]. In this method one picks at random one of the endsites (i.e. a site having only one neighbour in the tree) of a tree and deletes it. Subsequently, one selects at random any of the perimeter sites of the tree and occupies it. If this doesn't create a loop, one has generated a new tree. Otherwise one goes back to the original tree and repeats the procedure. Caracciolo and Glaus [203] introduced a method similar to the Berretti–Sokal method for SAWs mentioned in chapter 3. Suppose at time step t one has a particular tree configuration. Then one picks at random a site i in the tree and either tries to delete the site with probability p (which is possible if the site is a perimeter site), or tries to add an extra bond at that site with probability $1 - p$, going in the direction of one of the neighbours of i (which is possible if that neighbour isn't already occupied). If this procedure fails, the starting configuration is kept as the 'new' configuration at time $t + 1$. With this method one can also determine the connective constant of trees and the exponent θ. One finds $\nu = 0.640 \pm 0.004$ and $\theta = 1.001 \pm 0.024$ for trees in $d = 2$. The estimate of the connective constant $\mu = 5.1434 \pm 0.0013$ is in agreement with the value from exact enumerations. In $d = 3$, the values are $\nu =$

0.495 ± 0.009, $\theta = 1.501\pm0.012$ and $\mu = 10.5439$. The Monte Carlo methods discussed so far use local moves. A variant of the pivot algorithm for trees has been introduced by Janse van Rensburg and Madras [204]. In this algorithm one randomly picks an edge of the tree and deletes it. This breaks the tree into two parts. As in the pivot algorithm, one performs one of the symmetry operations of the lattice on the smaller of the two parts. Then one picks two vertices at random, one in each part, and translates the parts so that the two chosen vertices become nearest neighbours. If this introduces no loops or crossings, a new tree has been generated. It was proven by Janse van Rensburg and Madras that this algorithm is ergodic. With this algorithm several properties of trees have been studied in dimensions ranging between $d = 2$ and $d = 9$. We quote here only the results for ν. They are $\nu = 0.6374\pm0.0071$ ($d = 2$), $\nu = 0.4960\pm0.0031$ ($d = 3$), $\nu = 0.4200\pm0.0079$ ($d = 4$), $\nu = 0.2651 \pm 0.0010$ ($d = 8$) and $\nu = 0.2560 \pm 0.0010$ ($d = 9$). In summary, we can conclude that the Parisi–Sourlas relation and the exponents in table 9.2 seem to be well confirmed by the available numerical evidence.

$\sim$

We conclude this section by a brief discussion of 'c-animals'. By this we mean any lattice animal with c cycles (or loops). Thus, trees are 0-animals. Let us denote by $g_v(c)$ the number of c-animals with v vertices. In analogy to (9.8) we expect that

$$g_v(c) \sim \mu_v(c)^v v^{-\theta(c)} \qquad (9.14)$$

Some remarkable results are known about $\mu_v(c)$ and $\theta(c)$. First of all $\mu_v(c)$ does not depend on c, and can be shown to be strictly less than μ_v, the connective constant of all site animals on the same lattice [205]. Secondly, it was proven by Soteros and Whittington [206] that

$$\theta(c) = \theta(0) - c \qquad (9.15)$$

a result which was first conjectured on the basis of exact enumerations [207]. As discussed above, within numerical accuracy one has $\theta = \theta(0)$. Relations similar to (9.15) will be encountered in the study of knots in SAPs (chapter 10) and in our study of self avoiding surfaces in chapter 11. Furthermore we will encounter c-animals in section 9.4 when we discuss the collapse of branched polymers.

9.3 Vesicles in $d = 2$

In this section we will discuss a model which, depending on a certain parameter, shows either the behaviour of a SAP or the behaviour of a branched polymer. The model was introduced to study the behaviour of vesicles. It is inspired by the membranes of biological cells. Cell membranes usually consist of a lipid bilayer in which proteins are immersed. These proteins play a crucial role in the selective transport of molecules through the cell membrane. Vesicles are simplified membranes which don't contain proteins and which consist of one or two layers of lipid that form a closed surface. To model real vesicles one has to work of course in three dimensions and model them with, for example, a self avoiding surface (see chapter 11). In this section we will discuss a vesicle model in two dimensions, which, even though it is less realistic, has the advantage that it is simpler and that several exact results can be obtained for it. These results can then be used as a first step in understanding vesicles in $d = 3$. Furthermore, as we will see in the next section, vesicles are also helpful in understanding the collapse of branched polymers.

To model vesicles, Leibler *et al.*[208] introduced a model which takes into account both a bending rigidity κ and pressure difference $\Delta p = p_{\text{int}} - p_{\text{ext}}$ between the inside pressure p_{int} and the pressure p_{ext} outside the vesicle. Such a difference can be caused, say, by a difference in osmotic pressure. The vesicle itself is modelled as a SAP. As mentioned in the first chapter, bending rigidity is an irrelevant quantity for linear polymers, so in this section we will neglect the effect of κ. As we will see in chapter 11, for surfaces bending rigidity can be of importance. Thus, in the vesicle model in $d = 2$, we will only retain the term in Δp (interesting effects do occur when one takes rigidity into account; for these we refer the reader to the literature [209]). The ensemble in which Δp is fixed is called the *stress ensemble*. Consider a SAP of N steps which has an enclosed area A. The weight of such a SAP in the stress ensemble is given as $\exp{(A\Delta p)}$. The appropriate partition function to study is therefore

$$Z_N^v(\Delta p) = \sum_A q_N(A) \exp{(A\Delta p)} \qquad (9.16)$$

where $q_N(A)$ gives the number of N-step polygons of area A. As

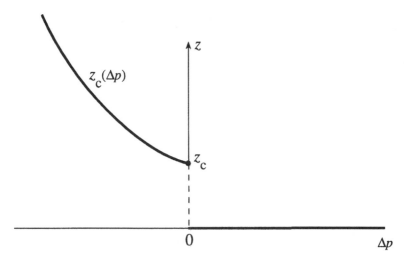

Fig. 9.4. Phase diagram for vesicles in $d = 2$. The broken line is a line of first order transitions.

usual we are interested in the free energy

$$f^v(\Delta p) = \lim_{N \to \infty} \frac{1}{N} \log Z_N^v(\Delta p) \qquad (9.17)$$

The properties of (9.16) and (9.17) have been studied by several authors, but the most extensive work is that of Fisher *et al.* [210]. We will to a large extent follow that reference in our discussion of vesicles. From the discussions in previous chapters, it is obvious that the grand canonical partition function of this model becomes critical at $z_c(\Delta p) = \exp[-f^v(\Delta p)]$. For $\Delta p = 0$, we of course recover the SAP result $z_c(0) = 1/\mu$. In the remainder of this section we will discuss the behaviour of $z_c(\Delta p)$ and the associated critical exponents. In figure 9.4 we have summarised the phase diagram. First, consider the case $\Delta p > 0$. When N is a multiple of 4 a SAP of maximal area is a square of area $A = N^2/16$. This SAP is unique. When N is not a multiple of 4, the largest area can be achieved by a rectangular SAP of sides $(N - 2)/4$ by $(N + 2)/4$ with a total area $(N^2 - 4)/16$. There are two such SAPs. In general therefore, $A \leq N^2/16$ and for $\Delta p > 0$ we therefore have

$$Z_N^v(\Delta p) \leq q_N \exp(\Delta p \, N^2/16) \qquad (9.18)$$

Moreover, taking only the term with the largest area in (9.16) we get a lower bound for $Z_N^v(\Delta p)$ proportional to $\exp(\Delta p N^2/16)$.

These bounds then imply that for $\Delta p > 0$, $f^v(\Delta p)$ diverges and therefore we obtain in that regime $z_c(\Delta p) = 0$. When $\Delta p < 0$ we have the obvious upper bound $Z_N^v(\Delta p) \leq q_N$. Concatenation arguments for SAPs (chapter 2) can be immediately extended to vesicles. If we concatenate any SAP of N_1 bonds and area A_1 with another SAP of N_2 bonds and area A_2 we get a polygon of $N_1 + N_2$ bonds with an area $A_1 + A_2 + 1$. This leads to the inequality

$$q_{N_1+N_2}(A) \geq \sum_{A_1} q_{N_1}(A_1) q_{N_2}(A - A_1 - 1)$$

Inserting this inequality in (9.16) we get for $Z_N^v(\Delta p)$

$$Z_N^v(\Delta p) \exp(\Delta p) \geq [Z_{N_1}^v(\Delta p) \exp(\Delta p)][Z_{N_2}^v(\Delta p) \exp(\Delta p)] \quad (9.19)$$

Together with the upper bound on Z_N^v this implies the existence of the free energy in the case $\Delta p < 0$. Furthermore, by using convex analysis and equation (9.19), it can be shown [210] that $z_c(\Delta p)$ is continuous for $\Delta p < 0$ and is left-continuous for $\Delta p \to 0$. Finally, it is possible to show that the line $\Delta p = 0$, for $z < z_c(0)$, is a line of essential singularities of the grand canonical partition function [210].

We now want to investigate the critical properties of the regimes where $\Delta p \neq 0$. The average area $\langle A_N(\Delta p) \rangle$ inside the vesicle can be obtained by making a suitable derivative of $Z_N^v(\Delta p)$

$$\langle A_N(\Delta p) \rangle = \frac{d \log Z_N^v(\Delta p)}{d(\Delta p)} \quad (9.20)$$

From the discussion above, it follows then that for $\Delta p > 0$, $A_N(\Delta p) \sim N^2$. More precisely, it is not difficult to see that

$$\lim_{N \to \infty} A_N(\Delta p)/N^2 = 1/16$$

For $\Delta p < 0$, the fact that f_N^v is finite implies that the average area grows proportional with N. For these reasons we speak of *inflated* vesicles when $\Delta p > 0$, and *deflated* vesicles in the other situation. Suppose then that, in a grand canonical description, we keep z fixed at some value less than $z_c(0)$ and that we increase Δp from a negative to a positive value. When we cross the line of essential singularities at $\Delta p = 0$ the average area makes a discontinuous jump from a finite value to an infinite value. When we do the same at some $z \geq z_c(0)$ the average area increases continuously until it reaches an infinite value along the line $z_c(\Delta p)$.

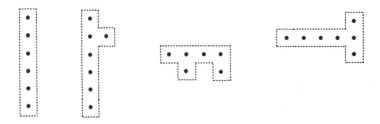

Fig. 9.5. Some vesicles of minimal area with $N = 14, A = 6$.

We are now ready to explain why for $\Delta p < 0$ it is believed that
the critical behaviour of vesicles is that of branched polymers.
The minimal area inside a SAP of length N is obtained when
$A = (N - 2)/2$. This is the case when the SAP is a rectangle of
$(N - 2)/2$ by 1. But there are other configurations having this
minimal area. See figure 9.5 for some examples. As indicated in
that figure, by going to the dual lattice we can associate with
such a vesicle a lattice animal, which owing to the minimum area
requirement has to be a tree. (However, not all trees correspond
to a minimal area vesicle.) Let the number of such trees with
N vertices be $\hat{t}_N$. By concatenation arguments one can show the
existence of a connective constant μ_{mat} for these minimal area
trees (mat). In this way, we obtain bounds for (9.16) in the deflated
vesicle regime

$$q_N \exp\left(\Delta p \frac{N-2}{2}\right) \geq Z_N^v(\Delta p) \geq \hat{t}_N \exp\left(\Delta p \frac{N-2}{2}\right)$$

From this we get bounds for $f^v(\Delta p)$

$$\log \mu + \frac{1}{2}\Delta p \geq f^v(\Delta p) \geq \log \mu_{\text{mat}} + \frac{1}{2}\Delta p$$

These inequalities show that for $\Delta p \to -\infty$,$f^v(\Delta p)/\Delta p$ should
approach $1/2$. In that regime the partition function is dominated
by the minimal area trees. This fact leads us to suspect that for
$\Delta p \to -\infty$ the critical behaviour of (9.17) is that of trees, i.e.
that of branched polymers. Since there are no phase transitions
in the regime $\Delta p < 0$, this should in fact be the behaviour in the
whole deflated regime. Indeed, all numerical evidence points to
the fact that in the deflated regime vesicles behave like branched
polymers; the vesicles then adopt a double stranded shape in which
each strand becomes a branch of a tree-like structure. For example,

Monte Carlo simulation studies of vesicles in the regime $\Delta p \to -\infty$ give an exponent $\nu = 0.65 \pm 0.04$, nicely consistent with the known ν-exponent for BPs in $d = 2$ [208].

A lot of attention has been paid to the crossover behaviour from SAP to BP behaviour which occurs at $\Delta p = 0$. Consider therefore the average radius of gyration. We expect that near $\Delta p = 0$ and for large N it is of the form

$$\langle R_N^2(\Delta p) \rangle \approx N^{2\nu} F(\Delta p N^\varphi) \qquad (9.21)$$

Here ν is of course the SAW exponent and φ is a crossover exponent. From the discussion so far it is obvious that $\varphi > 0$. The average area inside the vesicle (9.17) should obey a similar scaling law

$$\langle A_N(\Delta p) \rangle \approx N^{2\nu_A} G(\Delta p N^\varphi) \qquad (9.22)$$

where now ν_A is a new exponent which describes how the average area within a SAP scales with N. This exponent was first studied numerically by Enting and Guttmann [211]. These authors determined that within numerical accuracy $\nu_A = \nu$, which implies that for a SAP

$$\langle R_N^2 \rangle \sim \langle A_N \rangle$$

or, stated otherwise, the interior of a SAP is compact (not fractal). It was pointed out by Duplantier [212] that the fractal dimension D within the loops of the $O(n)$-model can be determined using the Coulomb gas. This calculation confirmed that for SAPs, D is indeed 2. Since in the vesicle model Δp couples to the area, it is natural that the exponent of N in the scaling laws (9.21)–(9.22) should be 2ν. Thus we arrive at the conclusion that the crossover exponent $y_{\Delta p} = \varphi/\nu$ at $\Delta p = 0$ equals 2. Indeed, if one takes this value of φ the scalings are very well satisfied numerically [208]. In chapter 11 we will encounter another argument that shows that $y_{\Delta p} = 2$ [213].

We conclude by mentioning that all arguments in this section about the vesicle phase diagram can be extended to the three-dimensional case, so that, at least qualitatively, the phase diagram of 'real' vesicles looks similar to that shown in figure 9.4 (as long as bending energies are neglected).

$$\sim$$

The vesicle model shows within one model both linear and branched polymer behaviour. This has led to further investigation of models which allow a transition between the two kinds of behaviour. For instance, one can study the crossover between SAWs and trees by giving a weight to the number of vertices in a tree which are of order one (i.e. which have one edge incident on them). If $t_v(E)$ is the number of trees with v vertices, E of which are of order one, one introduces the partition function

$$Z_v(x) = \sum_E t_v(E)x^E \tag{9.23}$$

When $x = 0$ only SAWs contribute to (9.23) whereas at $x = 1$ all trees get equal weight. Is there a transition in this model at some non-zero x_c? The answer is no, since x is a relevant variable at the SAW fixed point. This was pointed out in a paper by Camacho and Fisher [214] . When x is very small the sum (9.23) will (besides SAWs) mainly get contributions from trees with $E = 3$ which are nothing but the stars with three branches which we encountered in section 9.1. But since γ_3 is greater than the SAW-γ, there is a crossover to tree behaviour as soon as x becomes non-zero; the relevant crossover exponent is therefore $\gamma_3 - \gamma = 49/16 - 43/32 = 55/32$. For further discussion of this model, also including a bending rigidity, we refer to [214].

Another model showing both linear and branched polymer behaviour was introduced by Orlandini et $al.$ [215]. Theirs is a model for so called two-tolerant walks [216]. These are SAWs which can cross each edge of a lattice at most twice. Let the number of doubly visited edges be T. A suitable fugacity M gives a weight coupled to T. The model was studied with exact enumeration methods. When M is large, polymer configurations which have double stranded shape are favoured. These are not very different from those of vesicles in the extremely deflated regime, so we expect at large M a BP behaviour. A transition from linear to branched polymer behaviour was indeed found numerically near $M_c \approx 2.0$. In the high-M region exponents consistent with BP behaviour are found. This models differs from the previous models in the fact that there is a SAW phase for a whole region of parameter values. This can be understood as follows. For M slightly above 1, a small number of doubly visited edges are introduced in the model. In the language of the $O(n)$-model they can be described by adding a term $\sum_{\langle i,j \rangle} (\vec{s}_i \cdot \vec{s}_j)^2$ to the partition function. In a high temperature expansion such a term can be seen to lead to doubly visited edges. Correlation functions between two doubly visited edges decay with a power law involving the watermelon exponent x_4. This is obvious from the fact that at each end of such an edge two bonds must enter. From (3.5) one can then find that the RG eigenvalue $y_4 = 2 - x_4$ is irrelevant at the SAW fixed point. That is why the model of Orlandini *et al.* has a transition at some finite value of M.

9.4 Collapse and adsorption of branched polymers

We have seen in chapter 8 that the collapse of linear polymers is understood rather well by now and that it is generally believed that there is one universality class describing the θ-point in various models. For BP the collapse transition has been the focus of a lot of work in recent years [217, 218]. So far most attention has been given to the two-dimensional case. Since there are several different ways to count BPs, several collapse models can be introduced. In this section we will present evidence that there exist two universality classes for BP collapse separated by a third class where the collapse is described by percolation. So strangely enough, percolation plays a role in the collapse of both branched and linear polymers (a model which tries to combine both collapses in one model has been introduced in [219]).

Two main classes of branched polymer collapse have been introduced. They are

• *Cycle models* Here one considers the c-animals encountered at the end of section 9.2 and gives different weights to animals with different numbers of cycles. Increasing this weight increases the number of cycles which leads to a collapse.

• *Contact models* Here one works with a weak embedding of the animal and calls a contact a pair of nearest neighbour vertices of the animal which have no edge joining them. One then gives a weight to the number of these contacts. This model is more similar to the θ-model of SAWs.

We now study these models in more detail, beginning with the cycle model. Let $G_v(s)$ be the number of (strongly) embedded bond animals with v vertices and s edges. We want to study the partition function

$$Z_v^{\text{cycle}} = \sum_s G_v(s) y^s \qquad (9.24)$$

This is a cycle model since from Euler's relation we know that the number of cycles c is given by $s - v + 1$ so that (9.24) can be rewritten in terms of a cycle fugacity also. As usual this partition function is an exponential in v, and the associated grand canonical partition function $\sum_v Z_v^{\text{cycle}} x^v$ diverges at some $x_c(y)$. When y is small the critical exponents at $x_c(y)$ are those of the free branched polymers discussed in section 9.2. At some x^θ, y^θ the BP is at its θ-point. This model has been studied by various authors using exact enumerations and Monte Carlo methods which yielded mainly estimates of the crossover exponent at the θ-point in both $d = 2$ and $d = 3$. For example in $d = 3$, analysing a series determined by Sykes and Wilkinson [220], Chang and Shapir [221] arrived at a value of ϕ_θ close to 1. A transfer matrix study of the cycle model (9.24) was performed by Derrida and Herrmann [222]. These authors obtained accurate values for x^θ, y^θ to which we will return in a moment. They also obtained the following values for the exponents at the θ-point: $\nu_\theta = 0.5095 \pm 0.003$, $\phi_\theta = 0.657 \pm 0.025$.

Some years ago, Vanderzande [223] studied a simplified version of the cycle model which can be mapped onto the vesicle model and from which the critical exponents can hence be determined exactly. The idea is to study a subset of the bond animals, namely to consider only those animals without any internal holes (figure 9.6). Such animals are also sometimes called discs. To see the relation with vesicles, draw (on the dual lattice) a perimeter round the animal. This perimeter is a SAP and the area A inside this SAP is just equal to the number of vertices in the animal. Moreover, each step of the SAP corresponds to one perimeter bond of the animal. The number of perimeter bonds can in turn be related to v and s. This gives for the number N of steps in the SAP (on the square lattice)

$$N = 4v - 2s$$

Therefore the grand canonical weight of an animal $x^v y^s$ can be

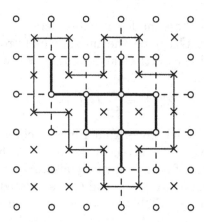

Fig. 9.6. Mapping a lattice animal without holes onto a vesicle. The open circles (crosses) indicate vertices of the original (dual) lattice. The thick full lines are the edges of the lattice animal, the thin full line those of the vesicle. Broken lines indicate perimeter edges of the animal.

rewritten for animals which are discs as

$$(xy^2)^A(y^{-1/2})^N \qquad (9.25)$$

But this is nothing other than the weight of a SAP in the vesicle model where $\exp(\Delta p) = xy^2$ and the step fugacity $z = y^{-1/2}$. We can then use our knowledge of vesicles to study the collapse of discs. The non-interacting BPs (i.e those with $y = 1$) are described by vesicles with $z = 1$. Increasing x one moves at fixed z in the deflated vesicle regime until one hits the critical line of the vesicles (figure 9.7). Thus the critical behaviour of the non-interacting lattice animals is that of vesicles in the deflated regime, a conclusion which is indeed consistent with what we found in the previous section. Increasing y means decreasing z in the vesicle model. This doesn't change the critical behaviour until y is equal to $z_c(0)^2$, when we hit the SAP fixed point. It is then obvious (figure 9.7) that the ν-exponent of the BP changes, so this must be the θ-point. At this point we immediately find from the known vesicle exponents that $\nu_\theta = 1/2$, $\phi_\theta = 2/3$. Notice that at this θ-point $x^\theta(y^\theta)^2 = 1$. Moreover $y^\theta = z_c(0)^2$. At even higher y-values, we see that the animal grand partition function diverges when we hit the line $\Delta p = 0$. Along this line the size of the animal makes a discon-

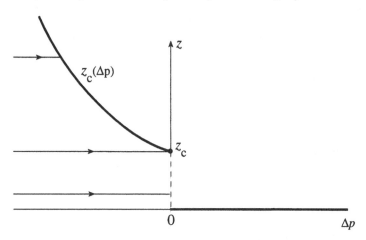

Fig. 9.7. Phase diagram of discs with cycle fugacity (see text).

tinuous jump from a finite to an infinite value, showing that in this regime the transition is indeed first order, and that the SAP point is a tricritical point. Furthermore, the location of the first order line is thus determined exactly and given as $x_c = y^{-2}$. Finally we can also estimate the exponent θ (see (9.8)) at the collapse point. The number of discs of v vertices can be related to the number of SAPs with a fixed area, which was determined numerically in the work of Enting and Guttmann [211]. From that work we get (in a somewhat confusing notation) $\theta_\theta = 2.0 \pm 0.1$. In conclusion then, we have determined the complete critical behaviour of a simplified cycle model of BP collapse. We can now ask: what is the effect of the holes? If we look at the numerical data of Derrida and Herrmann discussed above we see that the exponents found in that work are very close to those found in our model. Moreover, within numerical accuracy one still finds that even in the presence of holes the parameters x and y at the θ-point satisfy the relation $x^\theta (y^\theta)^2 = 1$ (in reference [224] arguments are given that this may not exactly be the case). Below we will introduce a more general model for BP collapse which can be described by a Potts model. That work will give further evidence that the holes are irrelevant and that the universality class of the θ-point in the cycle model is indeed the one found above.

∼

It is now time to turn to the contact model. Let $g_v(I)$ be the number of bond animals of v vertices and I contacts. The contact model is defined through the partition function

$$Z_v^{\text{contact}} = \sum_I g_v(I)\tau^I \qquad (9.26)$$

This model has been studied by exact enumerations [225]. Flesia *et al.* [226] have also introduced a more general model which contains the cycle and the contact model as special cases, and studied it using exact enumerations. In fact through a simple mapping this extended model can be related to a model studied earlier by Coniglio [227]. Coniglio showed how the general collapse model can be described by an extended Potts model, which he then studied using a real space RG approach. His model was more recently studied using the transfer matrix by Seno and Vanderzande [228]. In the rest of this section we will mainly follow the work of these authors, from which there is clear evidence for the existence of two different universality classes of branched polymer collapse. The arguments leading to this conclusion are somewhat technical and are described below (readers who are not interested in too many details can skip this part of the book, after looking briefly at figure 9.8).

In a general lattice animal we count the number of vertices v, the number of edges s and the number of contacts I. We want to study the grand canonical partition function

$$Z^{\text{gen}} = \sum_{v,s,I} g_v(s,I)x^v y^s \tau^I \qquad (9.27)$$

which for $\tau = 0$ gives the cycle model, and for $y = 1$ gives the contact model.

As mentioned above, Coniglio pointed out [227] that the grand partition function (9.27) can be obtained from the high temperature expansion of a Potts model. In this Potts model one adds extra terms to the Hamiltonian (6.13). In particular, the reduced Hamiltonian of the extended Potts model is

$$H_P^{\text{gen}} = K \sum_{\langle i,j \rangle} \delta_{\sigma_i \sigma_j} + L \sum_{\langle i,j \rangle} \delta_{\sigma_i 1} \delta_{\sigma_j 1} + H \sum_i \delta_{\sigma_i 1} \qquad (9.28)$$

One can perform a high temperature expansion of the partition function of this model using the techniques discussed in section 6.2. Terms in this expansion can be visualised as graphs that, like in the case of the Potts model itself, consist of a number of clusters on the lattice. When q goes to 1 the weight of a cluster becomes

$$(pw)^s \exp(-Hv)(1-p)^t \qquad (9.29)$$

Here t is the number of perimeter bonds of the cluster and

$$p = 1 - \exp(-K - L)$$
$$w = (\exp(K) - 1)/(\exp(J + L) - 1)$$

When $L = H = 0$, (9.28) reduces to the Potts model of section 6.2 which for $q = 1$ is critical (on the square lattice) for $K = K_c = \log 2$. In that case the weight of (9.29) reduces to that of bond percolation. We can transform the weight (9.29) into (9.27) by noticing that for any cluster (or animal)

$$4v = 2s + t + I$$

Inserting this into (9.29) we get expressions relating x, y and τ to the parameters of the Potts model

$$x = \exp(-H - 4(J + L))$$
$$y = [\exp(J) - 1]\exp(J + L) \qquad (9.30)$$
$$\tau = \exp(J + L)$$

The percolation point is therefore located at $y = \tau = 2, x = \frac{1}{16}$. It is furthermore known from Coulomb gas calculations that within the space of parameters of (9.28) the percolation point is fully repulsive; besides the thermal exponent $y_t = 3/4$ and the magnetic exponent $y_H = 91/48$ there is another exponent related to the symmetry breaking term in L, which for percolation is equal to y_t. In figure 9.8 we sketch the phase diagram of the generalised collapse model in the (y, τ)-plane. In this figure we have indicated the fully repulsive percolation fixed point and the projection of the line $L = H = 0$ along which our model coincides with bond percolation. We see that the lines corresponding to the cycle model and the contact model are on different sides of the percolation fixed point which is strong evidence that both collapse models are in different universality classes. What remains to be done is to determine the universality classes of these collapses. Seno and Vanderzande argue on the basis of the knowledge of the phase diagram of (9.28) for the case $q = 2$ [229], and the assumption that RG flow doesn't change qualitatively when one passes from $q = 2$ to $q = 1$, that the collapse transition of the cycle model is in

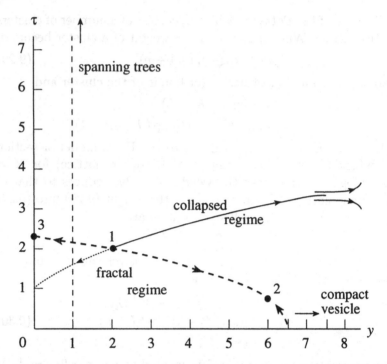

Fig. 9.8. Projection of the BP collapse phase diagram in the (y, τ)-plane. The arrows indicate renormalisation-group flows. Three critical points are indicated: (1) percolation point, (2) tricritical zero-state Potts point, and (3) Ising critical point. The full curve is the location of supercritical percolation, while the dotted curve indicates subcritical percolation. The broken curve is the line of θ-transitions. The line at $y = 1$ is the line along which the model coincides with the contact model of site animals. In the plane $\tau = 0$ the model coincides with the cycle model.

the universality class of the so called tricritical zero-Potts model (which is closely related to that of the critical $O(n = 0)$ model). This then gives the same exponents as those found from the simplified disc model. For the collapse of the contact model one can give arguments that indicate that it is in the Ising universality class implying that for that model $\nu_\theta = \phi_\theta = 8/15$. Extensive transfer matrix calculations of the general model give numerical estimates for the exponents which are indeed consistent with all these predictions. The θ-exponent at the collapse point of the contact model remains so far unknown.

What about the collapsed phase? Using the disc model and its relation to vesicles one can see that in the cycle model the BPs collapse into a compact state which has the properties of the interior of a SAP. For the contact model, on the other hand, it can be argued that they collapse into structures that look like spanning trees, which have the maximum number of contacts among all lattice animals. This follows also from the Potts model description. Indeed, for $H \to -\infty$ the Potts spins can no longer be in the state 1 and therefore the Hamiltonian (9.28) reduces to that of a $(q-1)$-state Potts model. In the case that interests us here $(q = 1)$, we therefore find that for $H \to -\infty$, a BP will have the criticality of spanning trees, which as the reader will remember from chapter 6 are the diagrams of the zero-state Potts model. It can be shown that for the contact model at $y = 1$, $x_c(\tau) \sim \tau^{-1}$ when $\tau \to \infty$ [217]. Along the critical line of the contact model one therefore has $x_c(\tau)\tau^4 \sim \tau^3$ for large τ. Comparing with (9.30) we realise that this is indeed the regime $H \to -\infty$.

The arguments we gave above on the phase diagram of the extended Potts model are quite general and should also hold in higher dimensions. In chapter 11 we will discuss the behaviour of vesicles in $d = 3$, which will give results for the cycle model for BP collapse. The results are $\nu_\theta = 1/2$, $\phi_\theta = 1$. On the other hand a very recent extensive Monte Carlo calculation on the collapse of trees, which should be in the universality class of the contact model, in $d = 2$ and $d = 3$ was performed by Madras and Janse van Rensburg. In $d = 2$ these authors determine the following exponent values: $\nu_\theta = 0.54 \pm 0.03$, $\phi_\theta = 0.579 \pm 0.022$. This value of φ is not consistent with the predicted Ising universality class exponent, but the value of ν is. In $d = 3$, the results are: $\nu_\theta = 0.40 \pm 0.005$ and $\phi_\theta = 0.655 \pm 0.024$. These numbers are precisely what can be found using the numerical values of Ising critical exponents in $d = 3$. In that case one should have $\nu_\theta = 1/y_H \simeq 0.40$ and $\phi_\theta = y_t/y_H \simeq 0.64$. In summary, then, there is quite some numerical evidence for the existence of two distinct universality classes of θ-transitions for branched polymers.

$\sim$

To conclude this chapter we briefly discuss the behaviour of branched polymers attached to a surface. It is not difficult to show that for branched polymers also there exists an adsorption transition. In that case one might be interested in determining the critical exponents at both the ordinary and the special (or adsorption) point. Let us first investigate the ordinary transition. Let t_v^s denote the number of trees having v vertices, at least one of which is in the surface. We expect that t_v^s has the form

$$t_v^s \sim \mu_t^v v^{-\theta_s}$$ (9.31)

Similar relations of course hold for bond and site animals. De'Bell *et al.* [231] showed that at the ordinary transition

$$\theta_s = \theta + 1 \qquad \text{ordinary point}$$ (9.32)

This results holds in any dimension and for all lattices. It is essentially a consequence of the fact that every bulk animal can be attached to a surface.

For the special point, no such rigorous results exist. Recently, Janssen and Lyssy [232] extended the work of Parisi–Sourlas to the case that a surface is present. This relates branched polymers attached to a surface in d dimensions to a semi-infinite Ising model in an imaginary field in $d - 2$. Besides the bulk field H it is now necessary to include a surface field H_s which can give rise to a special transition (see also section 5.1). Several interesting results were obtained by this approach. Firstly, from an exact study of the Ising model in an imaginary field in $d = 1$, the result $\phi_s = 1/2$ could be obtained for branched polymer adsorption in $d = 3$. The same value of the exponent was found in a study using conformal invariance for the Ising model in $d = 2$. A mean field approach and an ϵ-expansion gave the same result for ϕ_s. This led Janssen and Lyssy to suggest that this value for ϕ_s is superuniversal. Indeed, in chapter 5 we saw that the same value may also hold for linear polymers. Finally, we mention that a transfer matrix study of BP adsorption in $d = 2$ by de Queiroz [233] gives the estimate $\phi_s = 0.505 \pm 0.015$, again consistent with the superuniversality of this exponent. It remains an interesting challenge to find a proof for such a result. We remind the reader that the universality doesn't seem to hold at the θ-point of linear polymers. Furthermore it must be said that numerical estimates of ϕ_s in $d = 3$ generally give a higher value [234]. But then one has to admit that it can be quite difficult to estimate ϕ_s reliably. Janssen and Lyssy also

Table 9.3. *Overview of known, conjectured and numerical exponents for BP in $d = 2$*

	ν	θ	ϕ	θ_s(ordinary)	θ_s(special)	ϕ_s
good solvent	0.64	1		2	0.86	1/2
θ-point(I)	1/2	2	2/3	?	?	1/3
θ-point(II)	8/15	?	8/15	?	-	-

determined the exponent θ_s at the special point in general d. Their result is

$$\theta_s = \frac{d-3}{d-2}(\theta - 1) + (2 - \phi_s) \qquad \text{special point} \qquad (9.33)$$

Using (9.12) and (9.13) and assuming superuniversality for ϕ_s we can simplify this to

$$\theta_s = (d - 3)\nu + \frac{3}{2}$$

a result which also makes sense in $d = 2$. No numerical verifications of these results are known.

In their work on the collapse of BP, Seno and Vanderzande [228] also make some predictions for surface exponents at the θ-point, which are however very conjectural and have received no independent confirmation yet. We mention that at the θ-point of the cycle model, there is a prediction for $\phi_s = 1/3$, whereas there are arguments that there is no adsorption transition at or below the θ-point in the contact model.

Finally, surface exponents for BPs of fixed topology have been determined by Duplantier, as an extension of his work for bulk BPs [87]. We refer the reader to the original literature for the relevant results.

This ends our discussion of the critical behaviour of branched polymers. We see that in general the behaviour of these objects is rather well understood, though not all exponents are known exactly, not even in $d = 2$. Table 9.3 summarises the current status for the BP exponents in various situations in $d = 2$ (compare with table 8.1).

10

Polymer topology

From the discussion in previous chapters it has become clear that the critical behaviour of linear and branched polymers in a dilute regime is very well understood by now. This has led to a shift in interest to other than critical properties. In this chapter we want to investigate some geometrical and topological properties of polymers.

In our discussion so far we have assumed that the polymer is able to explore its whole phase space or equivalently that we can average over the whole set of N-step (random or self avoiding) walks. This may no longer be true when some kind of 'obstacle' hinders the motion of the polymers. This obstacle can be another polymer in the same solution or a real obstacle (such as a cylinder) immersed in the solvent. Most interestingly, in ring polymers the obstacle can be another part of the same polymer. This occurs when the polymer is knotted. As an example, the energy barrier which is present between states 1 and 2 of a polymer (see fig. 10.1) may be too high to be crossed at room temperatures. These considerations make it clear that topology plays an important role in the dynamics of polymers. In general the part of phase space

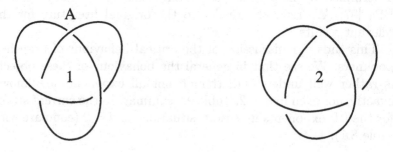

Fig. 10.1. Going from 1 to 2 requires a crossing of bonds at A.

176

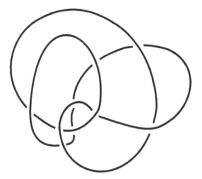

Fig. 10.2. An unknotted polymer can look very entangled.

which is available is that part in which all states of the polymer are topologically equivalent to the initial state.

The field of polymer topology was initiated by Sam Edwards [235] whilst investigating the entanglement between a polymer (described as a random walk) and an obstacle. Just as we have seen that a random walk can be related to the quantum mechanical study of a free particle in one dimension lower (chapter 1), a random walk in the presence of an obstacle can be related to the movement of a particle in a magnetic field (section 10.1). The study of the entanglement of a SAP with itself will lead to a study of knots in polymers (section 10.3). Knots can only occur in SAPs in $d = 3$. Their study requires first an introduction into the fascinating mathematical theory of knots. Even when a long polymer chain is unknotted it can give a very entangled impression (figure 10.2). This geometrical entanglement is measured by a quantity called the writhe, which we will study in section 10.4. Finally, we remark that to study topological properties essentially only two of the methods which we described in this book are useful. They are exact analytic results and Monte Carlo simulations. This is obvious since interesting topological effects can only occur in rather long chains (so there is no role for exact enumerations). Furthermore, knotting is a highly non-local property. This non-locality makes renormalisation group ideas, which use local rescalings, unusable. Some topological properties can however be written as a constraint (see section 10.1), and can then be rewritten as an extra term in a Lagrangian. If such a scheme could be extended to

Polymer topology

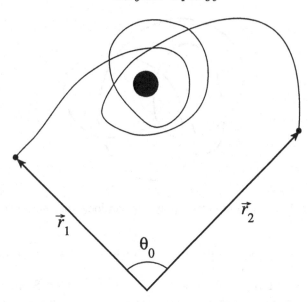

Fig. 10.3. A random walk near a disc-like obstacle.

knot properties, it might open a way for applications of the RG also in this domain of statistical mechanics.

10.1 A polymer chain near an obstacle

Consider a polymer chain (modelled as a random walk) whose ends are fixed, let's say at $\vec{r_1}$ and $\vec{r_2}$, and which can wind around an obstacle located at the origin (figure 10.3) [235]. A convenient topological invariant is the Gauss linking number G which counts the number of times the polymer winds around the obstacle. Let us say that we want to calculate the probability that an N-step random walk turns G times around the obstacle. The probability that the random walker goes from $\vec{r_1}$ to $\vec{r_2}$ is given by the path integral (1.12). We will take $t_0 = 0$ and work in $d = 2$. It is convenient to introduce polar coordinates (r, θ) to denote the position of the random walk. Then

$$\dot{\theta} = \frac{x\dot{y} - y\dot{x}}{x^2 + y^2}$$

The total angle covered by the walk in going between starting point and endpoint is

$$\Theta = \int_0^t \dot{\theta} dt \qquad (10.1)$$

Therefore the probability P_Θ (we don't write the obvious dependence on $\vec{r_1}$, $\vec{r_2}$ and t) that a random walker turns a total angle Θ is given by

$$P_\Theta = \int \mathcal{D}\vec{y}(t)\delta\left(\Theta - \int_0^t \frac{x\dot{y} - y\dot{x}}{x^2 + y^2} dt\right) \times \qquad (10.2)$$

$$\exp\left[-\frac{1}{4D}\int_0^t \dot{\vec{y}}^2(t')dt'\right]$$

Next we use the well known representation of the Dirac-δ

$$\delta(x) = \frac{1}{2\pi}\int_{-\infty}^{\infty} \exp\left(i\omega x\right) d\omega$$

and introduce the vector field $\vec{A}$

$$\vec{A} = \frac{1}{x^2 + y^2}\left(y\vec{e}_x - x\vec{e}_y\right)$$

to rewrite (10.2) as

$$P_\Theta = \frac{1}{2\pi}\int_{-\infty}^{\infty} e^{i\omega\Theta} d\omega \int \mathcal{D}\vec{y}(t) \times \qquad (10.3)$$

$$\exp\left[-\frac{1}{4D}\int_0^t (\dot{\vec{y}}^2(t') + i4\omega D\vec{A}\cdot\dot{\vec{y}})\, dt'\right]$$

The factor in the exponential

$$\dot{\vec{y}}^2 + i4D\omega\vec{A}\cdot\dot{\vec{y}}$$

is the Lagrangian (imaginary time) of a particle moving in a vector potential $\vec{A}$. The next step is then to go from the path integral approach to a Schrödinger equation. These steps are the reverse of the arguments going from the diffusion equation (1.9) to the path integral representation of its solution (1.12). The resulting Schrödinger equation, with the appropriate initial condition, can be solved by separation of variables in terms of the modified Bessel function I_p. We refer the interested reader to the original reference for the last steps in this derivation [235]. If Θ_0 denotes the angle between $\vec{r_1}$ and $\vec{r_2}$ (see figure 10.3), we can write $\Theta = \Theta_0 + 2\pi G$.

One then finally obtains an integral representation for the probability p_G that the polymer turns G times around the obstacle

$$p_G = \exp\left(-\frac{r_1^2 + r_2^2}{4Dt}\right) \int d\omega \exp\left[i\omega\left(\Theta_0 \pm 2\pi\left(G + \frac{1}{2} - \pi\right)\right)\right] I_{|\omega|}$$

The $+(-)$ refers to (counter) clockwise rotation around the obstacle. In this way we have determined exactly the entropy for a random walker with a given fixed value for a topological invariant (in this case G). We could go on to calculate the exponent ν for the subset of random walks having a fixed value of G. We would find that ν stays at $1/2$.

In section 10.3 we will ask similar questions for SAPs. The topological problem of a SAW with fixed endpoints near an obstacle is far less interesting since a SAW can only pass over or under the obstacle. For SAPs interesting topological effects are related to the occurrence of knots and links in the polygons.

To conclude this section, we mention that problems of random walks in the presence of two or more obstacles have been solved in recent years (see [236]).

10.2 Some elements of knot theory

Knot theory is one of the fascinating topics in mathematics. In this book we can only mention the most important concepts and definitions of this field in so far as they are necessary to understand polymer topology. For some nice introductions into this subject, see [238],[239].

Consider a closed curve in $\mathbb{R}^3$. In this book we will be mainly interested in the case that this closed curve is a SAP on some three-dimensional regular lattice. The curve is said to be *knotted* if it cannot be continuously deformed into a circle. The circle is sometimes called the *unknot*. Two knotted curves are called equivalent if one can be continuously deformed into the other (what we call here rather vaguely equivalent knots is defined more precisely through the concept of isotopy in topology). To discuss knots in SAPs it is convenient to project the curves into a plane. We are interested in 'regular' projections in which the projection has at most a finite number of points of intersection; intersection points are the images of exactly two points along the SAP (i.e. the points of intersection are double points); and vertices of the SAP are never mapped onto a double point. Numerically, a regular projection of a SAP is obtained by first rotating the SAP using a matrix with irrational entries. Then this rotated SAP is projected onto a plane. The figures 10.1 and 10.2 contain examples of such regular projections. At intersections one part of the arc goes over another part; it is common to speak of *underpasses* and *overpasses*.

The *crossing number* n_c of a curve is defined as the smallest number of crossings that one can have in any regular projection of that curve. The crossing number of the unknot is 0. The trefoil knot (figure 10.4a) has $n_c = 3$, while the figure-8 knot has $n_c = 4$. (figure 10.4b). A knot is called *composite* if it consists of two knots, neither of which is the unknot (figure 10.4c). When a knot is not a composite knot it is called a *prime* knot. All knots can be decomposed into prime knots. The operation of knot composition is usually denoted by a # sign.

Any knot can be transformed into the unknot by changing some of the crossings (in a non-continuous way) from under to over or vice versa. This number may change from projection to projection. Its minimal value is called the *unknotting number*.

We now come to a major question in knot theory, which is still not completely solved. How can we know that two knots, or two projections of a knot, are equivalent? This problem is tackled by the introduction of *knot invariants*, mathematical objects (numbers, polynomials...) which are assigned to a given curve or its projection and which are not changed under any continuous deformation of the knot. The simplest example of such a knot invariant is the crossing number, and the unknotting number is another such invariant. But these are not very fine discriminators. For example the left handed and the right handed trefoil knots have the same crossing number, yet one can prove that they are distinct knots. They are examples of chiral knots, i.e. knots which are not equivalent to their mirror image. In general, polynomials are better knot invariants. The first such polynomial was introduced in 1928 by Alexander [237]. The field of knot theory has been particularly active in the last ten years, partly owing to the discovery of several new classes of knot invariants. Many of these invariants are derived from exactly solvable models of statistical mechanics (chapter 3). We cannot pursue this fascinating link be-

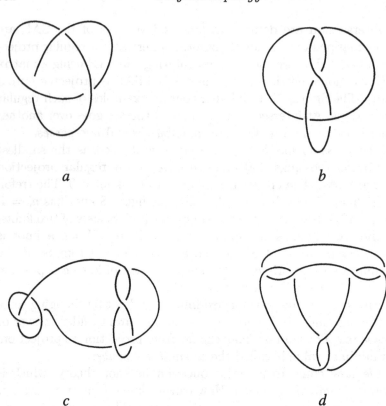

a   b

c   d

Fig. 10.4. Some knots and one link.

tween mathematics and statistical mechanics here.

$$\sim$$

For many calculational purposes the Alexander polynomial, denoted by $\Delta(t)$, is still used (see next section), so we will discuss here how it can be calculated. Usually, the Alexander polynomial is defined in terms of what is called a skein relation [238, 239], which relates the polynomial of a given projection to that of two other projections, one in which at a particular intersection under- and overpasses are interchanged and the other in which the intersection is removed. This definition is however not very useful in numerical calculations so we give another, equivalent definition.

The Alexander polynomial is calculated from a regular pro-

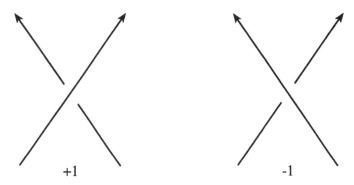

Fig. 10.5. Convention for crossings.

jection of the knot. If the crossing number is n_c, the Alexander polynomial is calculated from an n_c by n_c matrix A, which we shall call the Alexander matrix. To determine the elements in the matrix, one has to assign a direction to the projection. There are then two kinds of crossing (figure 10.5), which we count as positive and negative. The projection of the knot can be divided into n_c arcs, each of them going from one underpass to the next one. Each arc corresponds with one row in the Alexander matrix. If the arc overpassing at underpass k is arc l, the elements in the k-th row are given by

$$A_{kj} = (\delta_{kl} + \delta_{k+1,l})(\delta_{k+1,j} + \delta_{kj}) + (1 - \delta_{k+1,l})a(t)$$

where

$$a(t) = (\delta_{kj} - t\delta_{k+1,j} + (t+1)\delta_{lj}) \quad + \text{ crossings}$$
$$a(t) = (-t\delta_{kj} + \delta_{k+1,j} + (t-1)\delta_{lj}) \quad - \text{ crossings}$$

Finally one obtains the Alexander polynomial by deleting one row and one column from this matrix and calculating the determinant of the resulting matrix. For the unknot $\Delta(t) = 1$.

~

How many distinct prime knots are there with a fixed crossing number? These numbers have by now been counted up to $n_c = 15$ (for which there are 253,293 distinct knot types) [240]. There exist both lower [241] and upper bounds [242] on the number of distinct knots. Both bounds grow exponentially in n_c. This suggests that the number of distinct knots also grows as an exponential in n_c.

This could possibly be proved by a concatenation argument, but at this moment no such proof exists. If it could be found, it would be another case of statistical mechanics leading to an important advance in knot theory.

So far we have discussed the self entanglement of one single curve. In general one can consider several curves and discuss their linking (figure 10.4d). We will say little about links in this book. Two curves are linked if they cannot be continuously deformed into two separate circles. A topological invariant associated with a link is the so called linking number. To obtain this one first assigns an orientation to each curve in the link, and then one obtains a regular projection of the links. Thanks to the orientation the crossings can be given a sign, the convention for which is again the one shown in figure 10.5. The linking number is then $1/2$ of the sum of all the signs in the projection.

10.3 Knots in long polymers

We now come to a study of the occurrence of knots in polymers, and of the effect of these knots on polymer behaviour. We will limit ourselves to knots occurring in ring polymers (in particular to the occurrence of knots in SAPs). From a mathematical point of view, knots cannot occur in open curves. It is possible to give sense to a (quasi) knot for a SAW [243], but we will not go into that subject.

Figure 10.6 shows a trefoil knot in a SAP on the cubic lattice. The SAP has 24 bonds. In fact, it is possible to prove that this is the minimum number of bonds that one needs to build a trefoil knot on the cubic lattice [244]. One can ask in general how many bonds one needs before a certain type of knot can be built on a given lattice in $d = 3$. Or one can ask which fraction of all SAPs of more than 24 bonds contains a trefoil. The field studying these and other similar questions is sometimes called statistical knot theory. Other questions might be how the average crossing number of all SAPs of fixed N grows with N, and so on.

~

The first question which we ask ourselves is, what is the probability that a knot occurs in a SAP? Are knots rare or not? In fact, there existed a conjecture due to Delbruck [245] that every

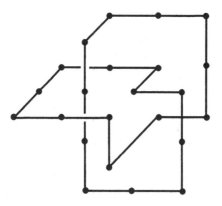

Fig. 10.6. A trefoil on the cubic lattice.

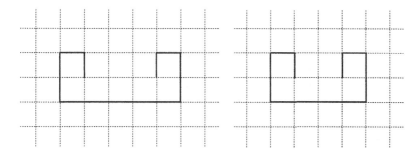

Fig. 10.7. The SAW on the left is a pattern, the one on the right is not.

sufficiently long polymer contains a knot. This result was later proven by Sumners and Whittington [246] and independently by Pippenger [247]. The proof of Sumners and Whittington relies on pattern theorems. By a *pattern* [17] one understands a part of a SAW which can occur in the 'middle' (here we mean not as the first nor as the last step) of a longer SAW. This implies that there should be a way into and out of the pattern (figure 10.7). Equivalently, a pattern is defined to be a SAW that can appear at (at least) three locations in a longer SAW. The SAW on the right of figure 10.7 can occur both at the end and at the beginning of a long SAW but nowhere else. In 1963, Kesten [248] proved a theorem about the occurrence of patterns, which is now known as Kesten's pattern theorem. Essentially, it says that for every pattern there exists a constant $c > 0$ such that the pattern occurs at least cN times on almost all N-step SAWs. The meaning of the 'almost all' is important; it means that the set of SAWs on which the pattern occurs less than cN times is exponentially rare. Stated more explicitly, Kesten's pattern theorem implies that if $\mathcal{P}$ is a pattern, the connective constant of SAWs that don't contain $\mathcal{P}$ is strictly smaller than that of all SAWs.

In their study, Sumners and Whittington introduced a pattern which when inserted into a larger SAP gave rise to a trefoil knot in the whole SAP. As a consequence of Kesten's pattern theorem, it then follows that the fraction of N-step SAPs that don't contain a trefoil goes down exponentially fast in N. This implies that unknotted SAPs are also exponentially few. If P_N^0 denotes the fraction of N-step SAPs that don't contain a knot, we have

$$P_N^0 = \exp\left[-\varepsilon_0 N + o(N)\right] \tag{10.4}$$

where $\varepsilon_0 > 0$. This is an important result. It shows that essentially all very long polymers are knotted. But it is also an asymptotic result. In fact for the range of N which is usually studied with, for example, a pivot algorithm, P_N^0 is still rather close to 1. This is reflected in the fact that ε_0 is small; as an example, on the FCC lattice $\varepsilon_0 = (7.6 \pm 0.9) \cdot 10^{-6}$ [249].

These numerical estimates are obtained as follows. First, for a given SAP one obtains a regular projection and an orientation is given to that projection. Then the location of crossings is determined and one has to determine their type. Finally, one calculates the Alexander polynomial in $t = -1$ ($|\Delta(-1)|$ is referred to as the order of a knot). This is usually sufficient to determine whether the SAP is knotted or not.

In most calculations performed so far, one considers polygons with lengths in the order of a few thousands. The knots appearing in these SAP are mostly rather simple and don't contain too many crossings ($n_c \le 10$ is a good estimate). Within this set of simple knot types, the order is a rather good knot invariant; there are only a few degeneracies. One could use more sophisticated knot polynomials to further discriminate between these knot types, but in general these new polynomials are computationally much harder to calculate.

The SAPs can be generated by one of the algorithms studied

in chapter 4. The pivot algorithm is efficient when one wants to sample over all knot types. If however one wants to restrict the dynamics to a set of topologically equivalent SAPs, the BFACF algorithm is used. Indeed, one can show that this local algorithm doesn't change the knot type [250]. On the other hand, this algorithm is rather slow, so one can use a mixture of both dynamics, rejecting SAPs of the wrong knot type.

With such numerical techniques, estimates like the one for ε_0 quoted above are obtained. In recent years extensive numerical work has been performed to determine the factors influencing ε_0. As an example, we mention first a study of the effect of solvent quality on ε_0. Janse van Rensburg and Whittington studied the knot probability as a function of temperature in the θ-model in $d = 3$ [249]. As can be expected intuitively, the knot probability goes up considerably when one is close to the θ-point. The exponent ε_0 accordingly becomes much larger. Vanderzande [251] studied the behaviour of the knot probability for polymers attached to a surface. For $\beta \to \infty$ in the polymer adsorption model of section 5.3, SAPs become completely adsorbed two-dimensional objects which cannot contain a knot. Extending the work of Sumners and Whittington, it can then be shown that ε_0 stays strictly positive for all positive temperatures. So even strongly adsorbed polymers will contain a knot for sure (if they are long enough).

These results thus show that the connective constant for unknotted polygons is strictly less than that for the set of all polygons. What about the effect of topology on critical exponents? Does α or ν depend on the type of knot present in the polymer? Let us first discuss the behaviour of the average radius of gyration where by average, we now mean average over a fixed knot type. An argument has been given by Stephen Quake [252] to show that the amplitude (but not the exponent ν) of the average radius of gyration should depend on crossing number. The argument is the following. A knotted SAP of N steps and crossing number n_c can essentially be seen as consisting of n_c loops each consisting of N/n_c monomers. These loops themselves are essentially free SAWs which are fixed at their 'endpoints', i.e. at the points where they are connected with other loops. Their average size R_{loop} therefore has the behaviour of a typical SAW and scales as

$$R_{\text{loop}} \sim \left(\frac{N}{n_c}\right)^{\nu}$$

in which ν takes on the value of the SAW in $d = 3$. Each loop will be able to move freely in a space $V_{\text{loop}} \sim R_{\text{loop}}^3$. The loops themselves can slide past each other and therefore the total volume in which the polymer moves is of the order $n_c V_{\text{loop}}$. Thus, the average radius of gyration R_g of the whole polymer is predicted to scale as

$$R_g(N, n_c) \sim n_c^{1/3-\nu} N^{\nu} \qquad (10.5)$$

If we insert the Flory value $\nu = 3/5$, the dependence on n_c has a power $-4/15$. Quake performed Monte Carlo simulations and did indeed find agreement with the predicted behaviour; ν is unchanged but the amplitude has a weak dependence on n_c. An independent study however disagrees partly with these conclusions. In their Monte Carlo work, Janse van Rensburg and Whittington [253] find that neither the exponent ν nor the amplitude depends on n_c, or in general on the knot type. Only the corrections to scaling are influenced by topology (E. Orlandini, personal communication). At this point, the matter is not settled and requires further investigation.

Recently, an interesting effect was found in the behaviour of the exponent α for knotted polymers. The behaviour is somewhat similar to what we discussed at the end of section 9.2 for c-animals. On the basis of a Monte Carlo investigation, Orlandini et al. [254] found that the exponent α depends on knot type in a very particular way. If n_p is the number of prime knots appearing in the knot decomposition of a particular SAP, the numerical work is consistent with the conjecture

$$\alpha(n_p) = \alpha(0) + n_p \qquad (10.6)$$

Moreover, the exponent $\alpha(0)$ of unknots is, within the numerical accuracy, the same as that for the set of all SAPs. It is possible that the relation (10.6) is just a consequence of 'rooting'; each prime knot is, in a very long chain, like one point on the SAP. If there are n_p prime factors, each of them is like a root that can be placed in (of the order of) N ways on the SAP. Therefore each prime factor increases α by 1. So far, a rigorous proof of this argument does not exist.

As a final topic in polymer topology we want to investigate how the 'complexity' of an average knot depends on N. When we consider longer and longer polymers, it is clear that they can contain more complex knots, where by 'complex' we mean knots with a higher crossing number or a higher unknotting number, for instance. A more precise definition of measures of knot complexity was introduced by Soteros *et al.* [182]. Any knot invariant M which is zero for the unknot and for which one can find a knot K such that

$$M(nK\#L) \geq nM(K) > 0, \quad \forall L \qquad (10.7)$$

(where L is any knot) is defined to be a good measure of knot complexity. Examples are the crossing number and the order of a knot. Using the pattern theorem, it is then possible to find a lower bound for the expected value of M. Indeed, from our discussion of the pattern theorem we know that for the trefoil there exists a real number c such that this pattern appears at least cN times in all but exponentially few SAPs. Then take an integer n_T such that $\lfloor (n_T - 1)c \rfloor = 0$ and $\lfloor n_T c \rfloor = 1$ where $\lfloor x \rfloor$ gives the largest integer smaller than or equal to x. So, for $N > n_T$ we can split the set of SAPs into a set W_1 of those SAPs which contain less than cN trefoils, and a set W_2 of SAPs which contain more than cN trefoils. Then, we have for the average value of the complexity $\langle M(N) \rangle$

$$\langle M(N) \rangle \quad \geq \quad \frac{1}{q_N} \sum_{W_2} M(W_2)$$

$$\geq \quad \frac{q'_N}{q_N} \lfloor Nc \rfloor M(T) \qquad (10.8)$$

where $M(T)$ is the value of M for the trefoil and we have used the relation (10.7). The factor q'_N is the cardinality of the set W_2. Finally, using the pattern theorem we get that for all but exponentially few SAPs

$$\langle M(N) \rangle \quad > \quad M(T) \left(\frac{N}{n_T} - 1 \right) \qquad (10.9)$$

The lower bound (10.9) suggests that $\langle M(N) \rangle$ may grow with a power of N

$$\langle M(N) \rangle \quad \sim \quad N^\tau \qquad (10.10)$$

Numerical work suggests that τ is very close to its lower bound. Orlandini *et al.* measured the average crossing number in simulations (using the pivot algorithm) of polygons with N up to 1500 [255]. These authors estimate $\tau = 1.122 \pm 0.005$.

10.4 The writhing of polymers

To conclude this chapter, we turn to a brief study of the writhe of a SAW or SAP. This geometrical quantity is known to be of great importance in the study of double stranded polymers such as DNA [256]. But it can also be introduced for a curve C such as a SAP. A convenient definition is the following. Give an orientation to C and project it in an plane perpendicular to any unit vector $\vec{n}$. In general the projection will contain (signed) crossings (figure 10.5). The writhe of that particular projection is defined as the sum of the signs of the crossings. Finally, the writhe of C is determined by averaging over all projection directions $\vec{n}$.

This definition of the writhe is hard to implement in a numerical calculation. Fortunately, for SAPs in $\mathbb{Z}^3$ a considerable simplification occurs, thanks to a theorem of Lacher and Sumners [257]. These authors have shown that the writhe can be calculated as the average of four linking numbers. To obtain these linking numbers one generates four new curves $C_1, \ldots, C_4$ by translating the original curve in four mutually non-antipodal directions (if $\vec{e}_1, \vec{e}_2, \vec{e}_3$ are the unit vectors of $\mathbb{Z}^3$ these four directions can be $(\vec{e}_1 + \vec{e}_2 + \vec{e}_3)/2, (-\vec{e}_1 + \vec{e}_2 + \vec{e}_3)/2, (-\vec{e}_1 - \vec{e}_2 + \vec{e}_3)/2$ and $\vec{e}_1 - \vec{e}_2 + \vec{e}_3)/2)$. Then, according to the theorem of Lacher and Sumners, the writhe is given by the average of the linking number of C with each C_i for $i = 1, \ldots, 4$. From this 'operational' definition of the writhe it is obvious that the writhe of a planar curve is zero. The writhe is therefore a measure of how twisted in space the curve C is. Moreover, the writhe of a curve which has a plane of symmetry is also zero. This can be seen by choosing the four translation directions in a symmetric way with respect to that plane. Therefore the writhe is also a quantity that is sensitive to the chirality of a curve.

A moment of reflection shows that for each SAP there exists another SAP for which the writhe W has exactly the opposite value. Therefore, the distribution of W is symmetric with respect to $W = 0$. In the following, we will therefore investigate the behaviour of $\langle |W| \rangle$.

Using the Lacher–Sumners theorem in combination with the pattern theorem, Janse van Rensburg *et al.* [258] were able to give a lower bound to the growth of $\langle |Wr| \rangle$ with the number of steps N in a polygon. The result is

$$\langle |Wr| \rangle \quad > \quad A\sqrt{N} \qquad (10.11)$$

As in the case of the knot complexity, this result suggests that the average value of the writhe grows with a power ω of N. Numerically, it is found that $\omega = 0.522 \pm 0.004$ [258]. This value is slightly larger then the lower bound. It is one of the current challenges to determine whether ω is really different from $1/2$ (for some work on the writhe of random walks, using field theoretic methods, see [259]). Another question is whether this exponent is universal and whether one can imagine situations in which it takes on a different value. For example, one can expect that in a collapsed phase ω should increase. Numerical work suggests however that at least for temperatures slightly below the θ-point, the change in ω is very small [255].

In figure 10.8 we show the distribution of the writhe for a set of SAPs generated with the pivot algorithm. This curve looks like a Gaussian whose width grows with N (whether it is really a Gaussian was not investigated in [258], but arguments that it should be Gaussian can be found in [260]). Maybe more interesting is the distribution of the writhe when we fix the knot type. An example is given in figure 10.9 for trefoils. These are chiral knots and we see nice evidence of a double peak structure in the distribution of the writhe. The relation between this distribution and the chirality of the knot has to be further investigated.

Polymer topology

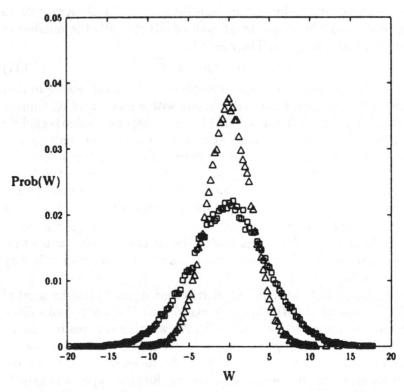

Fig. 10.8. Distribution of the writhe for a set of polygons of length $N = 400$ (triangles) and $N = 1100$ (squares) (from reference [258]).

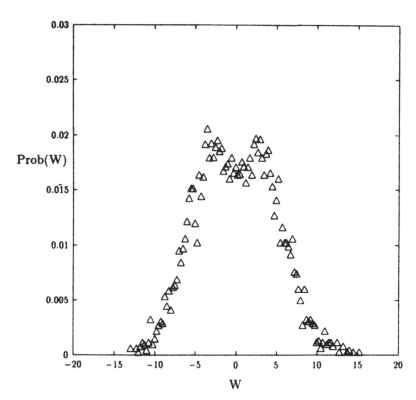

Fig. 10.9. Distribution of the writhe for trefoils with $N = 400$. The calculations were performed at a temperature close to the θ-point in $d = 3$ (from reference [258]).

11

Self avoiding surfaces

In this chapter we study self avoiding surfaces on a lattice. These surfaces are not immediately relevant for the study of polymers, although they could be of interest in the study of β-sheet polymers, which are important building blocks in proteins. Surfaces are used as models in the study of membranes or interfaces. Moreover, as we will see below, they allow the generalisation of vesicles to $d = 3$. The reason why we discuss surfaces in this book is mainly to show how the methods introduced in the statistical mechanics of polymers can be used in the study of other, but related, problems.

Several kinds of surfaces have been introduced in the literature [261]. A distinction has to be made between surfaces which are models of polymerised membranes and those which describe liquid membranes. In the first case, the number of nearest neighbours of a given monomer is fixed. An interesting model is that of so called tethered surfaces introduced by Kantor, Kardar and Nelson [262]. For liquid surfaces, on the other hand, the number of neighbours is not fixed. In this chapter, we will limit ourselves to a study of a lattice model of liquid surfaces, the 'plaquette' surfaces.

The critical behaviour of these surfaces is closely related to that of branched polymers. This is one of the many relations between surfaces and polymers.

~

Let us begin by defining the objects which we will study in this chapter and which will be referred to as plaquette surfaces or as *self avoiding surfaces* (SAS) [263]. An example of such a surface is shown in figure 11.1. The surface is built out of plaquettes of the cubic lattice. The number of plaquettes will be denoted by N, and plays a similar role to the number of monomers for a polymer. Each plaquette can occur at most once in a SAS and at most two plaquettes are allowed to meet at an edge. In this chapter we will

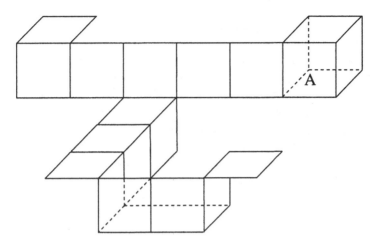

Fig. 11.1. A plaquette surface. The plaquette A does not belong to the surface.

give an overview of the properties of these surfaces and we will ask questions similar to those asked for polymers; how many surfaces there are of a given area N, how the 'size' of surface grows with N, what the effect of topology is, and so on.

Surfaces can be classified in several ways [264]. Firstly, they can be closed or open. Secondly, we can distinguish between orientable surfaces (such as a sphere) or the non-orientable ones. Of these latter, the Möbius strip is the most famous example. In the rest of this chapter we will mainly deal with orientable and closed surfaces. For these there exist an important topological invariant which is the genus H of the surface. It counts the number of handles on the surface. The genus can also be related to the average curvature of the surface.

11.1 Surfaces and branched polymers

In this section we start to develop the theory of SAS in a way similar to that for polymers in previous chapters. First, we can ask about entropic properties, i.e. what is the number S_N of surfaces of area N (counted per lattice site). We expect that this quantity has the usual behaviour

$$S_N \sim \mu_\square^N N^{-\theta} \tag{11.1}$$

where $\mu_\square$ is a connective constant for plaquette surfaces and where we have already anticipated the fact that the behaviour of surfaces will resemble that of branched polymers by using the exponent θ. (Needless to say, as usual the exponent ν describes the behaviour of the radius of gyration with N. The fractal dimension D of the surface is again given by $1/\nu$.) What is the evidence for this behaviour? Firstly, we remark that like in the case of polymers, one can introduce concatenation arguments which show rigorously that the leading term in S_N is indeed exponential [265]. Secondly, an extension of de Gennes' theorem to the case of surfaces can be given. Indeed, just as SAWs can be obtained from the limit as $n \to 0$ of a spin model, the $O(n)$-model, we can relate SAS to the limit as $n \to 0$ of a lattice gauge theory. This will lead to a connection between the grand partition function of the SAS-problem and the partition function of the lattice gauge theory [266]. An inverse Laplace transform then leads to the result (11.1). Moreover, one again has the relation $\mu_\square = 1/K_c$, where K_c is the critical temperature of the lattice gauge theory.

Lattice gauge theories [267] were introduced in high energy physics as lattice versions of field theories which have a local symmetry (such as QED and QCD). Just as polymer-like graphs occur in the high T-expansion of spin models, SAS occur naturally in the weak coupling expansion of lattice gauge theories. In fact, this was one of the original motivations for studying models of surfaces on a lattice.

We start by introducing the simplest lattice gauge theory which is the Ising lattice gauge theory. In this model one has on each edge e of a hypercubic lattice an Ising variable s_e. Four spins along the boundary of a plaquette P interact with each other

$$H_P = K s_{P,1} s_{P,2} s_{P,3} s_{P,4}$$

where P, i denotes the i-th edge in plaquette P. The partition function of the Ising lattice gauge theory (ILGT) is then

$$Z_{\text{ILGT}} = \text{Tr} \prod_P \exp H_P \qquad (11.2)$$

This ILGT has a local symmetry (apart from the global spin reversal symmetry). When we change all the spins on edges that are incident on one particular vertex the weight $\exp H_P$ remains unchanged. In this way we have a model with a local spin inversion symmetry. The ILGT is known to be critical when $d \geq 3$ [267]. In fact, in $d = 3$ there is a duality linking the small K regime with the high K region. This duality fixes the location of the critical point.

In a similar way, it is possible to introduce a lattice gauge theory with an $O(n)$-symmetry. Therefore one introduces n-component spin variables $\vec{s}_e$ along the edges of a hypercubic lattice. The plaquette interaction is

$$H_{\mathrm{P}}(n) = K \sum_{\alpha=1}^{n} s_{P,1}^{\alpha} s_{P,2}^{\alpha} s_{P,3}^{\alpha} s_{P,4}^{\alpha} \qquad (11.3)$$

and the corresponding partition function can conveniently be written down à la Nienhuis

$$Z_{\mathrm{LGT}}(n) = \mathrm{Tr} \prod_{P} (1 + H_{\mathrm{P}}(n)) \qquad (11.4)$$

Working out the product and performing the trace using the averaging properties of $O(n)$-spins we quickly see that the leading term (in n) in the expansion is given by closed SAS in which along each edge two plaquettes (and thus two spins) come together. Graphs in which four plaquettes come together at one edge give higher order contributions in n. The surfaces which are generated in this way have unrestricted topology, i.e. they can have an arbitrary number of handles. Open surfaces can be related to the Wilson loop, which in LGT plays a role similar to that of correlation functions in spin systems [267]. We will not discuss that relation here.

In principle this relation between SAS and lattice gauge theories can be used to gain information on the critical behaviour of the surfaces, but at this moment much less is known about lattice gauge theories in $d = 3$ than about spin models in $d = 2$. No connective constant or θ-exponent can be derived from this relation.

Very few analytical results exist for SAS. A study of surfaces without the self avoidance constraint determined the ν-exponent in that case as $1/4$ [263]. The law of codimension additivity (chapter 2) then suggests that the upper critical dimension for surfaces is 8. As in the case of SAWs we can then try to built a theory of Flory type for the exponent ν. This was first done by Maritan and Stella [268]. These authors assumed that the repulsive term in the free energy is still determined by monomer–monomer contacts and is unmodified with respect to the linear polymer case. Secondly, they assumed that for surfaces the area (of some projection of the surface) has a Gaussian distribution (this is a natural extension of the linear case in which the length of the polymer has this distribution, see chapter 1), which determines the elastic part of the free energy. The total free energy is therefore of the form (compare with (2.13))

$$F_N(R) = A\frac{N^2}{R^d} + TB\frac{R^4}{N}$$

which leads to the following Flory estimate for the exponent ν

$$\begin{aligned}
\nu &= \frac{3}{4+d} \quad (d \le 8) \\
&= \frac{1}{4} \quad (d > 8)
\end{aligned} \tag{11.5}$$

In this way we recover the result $d_c = 8$. Notice that $\nu = 1/4$ is exactly the exponent of branched polymers above their critical dimension. In contrast to the linear polymer case, Flory theory is in disagreement with numerical results in $d = 3$. Numerical data show, as we will discuss now, that in $d = 3$ also self avoiding surfaces behave as branched polymers.

Few of the numerical approaches introduced to study SAWs or lattice animals can be applied to the study of SAS. In an early study, Redner [269] applied an exact enumeration technique and found exponents consistent with branched polymer behaviour in, for example, $d = 3$ (remember that for such polymers $\theta = 3/2$ and $\nu = 1/2$ in $d = 3$). This study did not restrict the boundaries or the topology of the surfaces, and could only reach up to surfaces built out of 10 plaquettes.

A large amount of work has gone into developing good Monte Carlo approaches for surfaces. These have been pioneered by Sterling and Greensite [270]. They considered closed surfaces with $H = 0$, which are topologically equivalent to spheres. The method works in the grand canonical ensemble where a fugacity z is given to each plaquette of the surface. As we know from our study of Monte Carlo techniques, such a method allows the determination of quantities such as $\mu_\square$ and θ. In the method, one scans the cubes of the cubic lattices and when one encounters a cube with

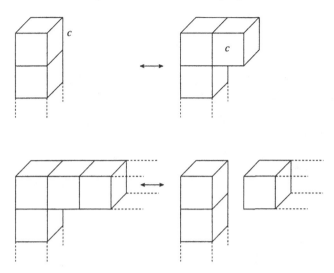

Fig. 11.2. Monte Carlo step for surfaces. On top we show an allowed move. The status of the faces of the cube indicated as C has been reversed. The move below can disconnect a surface and is disallowed.

at least one face in the surface, the status of each face of the cube is reversed; i.e. plaquettes that don't belong to the surface are included in the surface and vice versa. At the top of figure 11.2 we show an example. One then first checks whether the requirement of self avoidance is still obeyed and whether the topology has not been changed. For example, at the bottom of figure 11.2 we show a move that could disconnect a surface and is therefore not allowed. If the new surface satisfies all requirements it is accepted with a probability determined by the grand canonical weight. The original work of Sterling and Greensite studied a $10 \times 10 \times 10$ section of the cubic lattice and found exponents different from those of branched polymers. The algorithm of Sterling and Greensite was later modified by Glaus [271]. This allowed the scanning of a $20 \times 20 \times 20$ box of the cubic lattice. In this work good agreement with branched polymer behaviour was found; $\nu = 0.502 \pm 0.024$, and $\theta = 1.51 \pm 0.25$. The most extensive and clear results on the critical behaviour of SAS were obtained in the work of O'Connell *et al.* [272] who introduced several modifications of the Sterling and Greensite algorithm. These authors choose at random one of

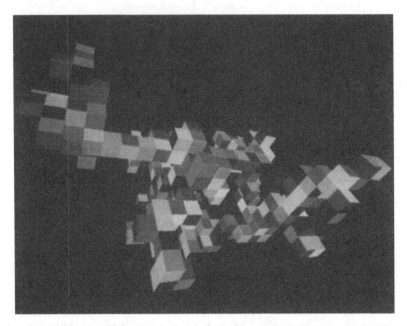

Fig. 11.3. A self avoiding random surface. The shading of the plaquettes has no meaning.

the plaquettes of the surface, and then choose one of the cubes
of the lattice that shares this plaquette. Then, the inside–outside
reversal is performed on that cube. In this way, some cubes which
have more faces in the surface are chosen with a higher probability; this effect has to be taken into account when calculating
averages. Another important ingredient of the work of O'Connell
et al. is the use of methods of sparse data-structures to store information about which cubes of the lattice touch the surface. We
refer the reader to the original reference for more details on these
techniques. In figure 11.3 we show a typical example of a surface
generated with this algorithm.

Using these techniques one can study surfaces with N in the
range $100-1000$. These studies have given rather strong numerical
evidence that the critical behaviour of SAS is that of branched

polymers; the estimates for the exponents are $\nu = 0.509 \pm 0.004$ and $\theta = 1.500 \pm 0.026$. These authors also estimate $\mu_\square = 1.729 \pm 0.036$. Further evidence for a branched polymer structure comes from a study of the ratio of the volume V enclosed by the surface, to the number of plaquettes N. From the data of O'Connell *et al.* it clearly follows that this ratio reaches a constant for large N. This implies that the fractal dimension of the surfaces equals $1/\nu \approx 2$. As we will see in the next section, this result is also important for the description of vesicles modelled by SAS. The fact that for large N, $V \sim N$ lends further support to a picture of large surfaces that consists of many long tubular filaments forming a tree-like structure. This picture is then consistent with branched polymer behaviour.

An alternative Monte Carlo method was introduced by Baumgärtner and Romero [273] and works for closed surfaces. In this method one places up spins in the centre of all the cubes which are within the closed surface and down spins in the cubes which are outside the surface. Hence, the surface plays the role of an interface. Then both random spin flips and random spin exchanges are performed. One has to investigate at each step whether the modified interface is still connected and of the desired topology. When one also takes care of keeping N fixed one has an algorithm that works in the canonical ensemble. The work of Baumgärtner and Romero confirmed the results of O'Connell *et al.* and led to the estimate $D = 1/\nu = 2.00 \pm 0.03$. No estimate of $\mu_\square$ or θ can be obtained with this method.

In a way, the results obtained from these simulations are disappointing since despite all efforts we find nothing new when studying SAS. We just recover the well known univerality class of branched polymers.

$$\sim$$

To conclude this section, we discuss briefly what is known about the adsorption and collapse of SAS. We have devoted a lot of attention to these phenomena for linear and branched polymers. As we will see much less is known about the same models for SAS.

Adsorption of closed SAS (with $g = 0$) has been studied numerically by Orlandini *et al.* [274]. Since closed SAS can be considered as vesicles with $\Delta p = 0$, their study is of relevance for the important phenomenon of vesicle adsorption. The model studied by

Orlandini *et al.* is an immediate generalisation of the adsorption
model for linear or branched polymers, in which we now give an
extra weight $(\exp \beta)$ to each plaquette in the surface. This model
is then studied using the Monte Carlo technique of O'Connell
et al.. Clear evidence is found for an adsorption transition at
a value of $\beta_c \approx 0.35$. At lower values of β, ν is estimated as
0.506 ± 0.008, whereas in the low temperature regime an exponent
of $\nu = 0.64 \pm 0.01$ is found. This is consistent with the BP picture
of SAS. Indeed, the low temperature ν-exponent is exactly that of
two-dimensional branched polymers. The crossover exponent for
this adsorption transition is found to be $\phi_s = 0.70 \pm 0.06$. This
value is not in agreement with the prediction of a superuniversal
crossover exponent for BP adsorption, which would give $\phi_s = 1/2$.
But the value is consistent with that found in other simulations of
BP adsorption. More interestingly, besides this usual adsorption
transition a second transition in the adsorbed state is found at
lower temperatures (at $\beta_c' \approx 1.8$). Above this β-value, the exponent ν is found to be close to $1/2$ indicating that the adsorbed
branched polymer has collapsed. That such a transition can occur can be understood by considering completely adsorbed SAS
(which have a thickness of 1). If such a SAS has a perimeter of
length t the number of plaquettes in the surface is $(N-t)/2$. Thus
the completely adsorbed SAS can increase the number of plaquettes in the surface by decreasing the length of its perimeter, and
thereby becoming more compact. The value of the ν-exponent at
this second transition is $\nu = 0.54 \pm 0.03$, which is, at least numerically, consistent with the exponent at the θ-point in the contact
model of BP collapse (chapter 9).

$$\sim$$

There exists very little work on the collapse of SAS. There is
no study of the usual θ-model extended to surfaces. The only
work which has been done so far is an extension of the model
of SAWs coupled to Ising vacancies mentioned in section 8.2. In
such a model one couples an n-component lattice gauge theory
to an Ising model on the dual lattice [275]. In particular a gauge
theory on a dodecahedral lattice is coupled to an Ising model
on the FCC lattice. In a high temperature expansion one then
obtains SAS which have to 'live' in a region of up spins. This is
rather similar to the θ'-model of Duplantier and Saleur where a

SAW was restricted to a certain subset of hexagons. The main difference is that in the θ'-model, the forbidden hexagons occur with an independent probability, whereas in the present model the Ising spins are correlated. Going through arguments similar to those of the θ'-model, one can then relate the properties of the surfaces to those of the hulls of Ising correlated percolation. When this percolation process is critical (which in $d = 3$ occurs at a temperature different from the Ising critical temperature), the fractal dimension of the surfaces becomes equal to the hull fractal dimension of critical Ising percolation. Unfortunately, at this moment there exist only very rough numerical estimates for that dimension [276]. Nevertheless, it is highly unlikely that this hull dimension would also be equal to 2. The surfaces which one can obtain in this model are therefore interesting candidates for surfaces that are not in the branched polymer universality class.

11.2 Vesicles in $d = 3$

We now turn back to the vesicle model which we have already studied in $d = 2$ in section 9.3. The vesicles are modelled as closed SAS with a given topology. The model also includes the pressure difference Δp which in $d = 3$ couples to the volume V inside the vesicle.

Most of the arguments which we gave for two-dimensional vesicles can be generalised to the three-dimensional case [277]. Thus, deflated vesicles are expected to behave like branched polymers, whereas inflated vesicles are always critical. Both regimes are separated by a line of essential singularities (see figure 9.4 again). We will focus on the behaviour of the crossover exponent ϕ at $\Delta p = 0$. As should be obvious from our discussion on vesicles in chapter 9, the crossover exponent describes the growth of the average volume with the total surface of the vesicle. We have discussed this property of surfaces in the previous section and we remind the reader that the exponent ϕ is numerically known to be 1. Therefore the RG eigenvalue $y_{\Delta p} = 1/\nu = 2$, for vesicles with the topology of a sphere. One expects that if the vesicle has a limited number of handles, this exponent will be unchanged. For vesicles with an arbitrary number of handles, an interesting argument has been given by Banavar et al. [278] that in that case $y_{\Delta p} = 3$. Their argument

204   *Self avoiding surfaces*

comes from a description of the vesicle model using an extension
of the lattice gauge theory (11.2). To define the gauge theory one
puts $O(n)$-spins on the edges of a (cubic) lattice and an Ising spin
σ_P at the centre of each plaquette. The Hamiltonian is then taken
to be

$$H = K \sum_P \sum_{\alpha=1}^N s_{P,1}^\alpha s_{P,2}^\alpha s_{P,3}^\alpha s_{P,4}^\alpha \sigma_P + u \sum_C \prod_{P \in C} \sigma_P \qquad (11.6)$$

where the second sum goes over all the cubes C of the lattice and
the product goes over the spins on the six faces of the cube. When
$u \to \infty$ the partition function will only contain terms with all
$\sigma_P = +1$; in this regime we recover the gauge theory of (11.3).
Let us now perform the usual high temperature expansion for the
term in K of this gauge theory, in which we first perform the trace
over the $O(n)$-spins. We consider in particular the limit $n \to 0$.
The diagrams that are obtained consist of SAS with an Ising spin
in each plaquette of the surface. The resulting partition function
can be written as

$$\begin{aligned} Z &= (\cosh u)^{N_C} \sum_{\sigma_P} \prod_C (1 + \tanh u \prod_{P \in C} \sigma_P) \times \\ &\quad \sum_{\text{SAS}} K^N \prod_{P_S} \sigma_P \end{aligned} \qquad (11.7)$$

where N_C is the number of cubes on the lattice and the final
product runs over the cubes inside the SAS. If we now perform the
sum over Ising spins only those terms contribute in which we take
a term in $\tanh u$ for each cube inside the surface. As a conclusion,
we find that our gauge theory can indeed describe vesicles in the
deflated regime and that

$$\Delta p = \log \tanh u \qquad (11.8)$$

We can therefore obtain the RG eigenvalue $y_{\Delta p}$ if we are able to
renormalise the Hamiltonian (11.6). Such a renormalisation can
indeed be performed because in the $n \to 0$ limit all RG contributions from the term in K vanish. The renormalisation of the
resulting theory involving only the term in u can be performed
by methods introduced by Kadanoff [279]. The result is simple:
$y_{\Delta p} = 3$. In fact, a completely similar analysis can be done in
$d = 2$, with the result $y_{\Delta p} = 2$ [213]. The $d = 2$ result is indeed the one which comes from the Coulomb gas calculation of

Duplantier, and is the one which is consistent with all numerical evidence (see section 9.3).

In conclusion, then, the work of Banavar *et al.* predicts that for SAS with an arbitrary number of handles (at $\Delta p = 0$)

$$\langle V \rangle \sim N^{3\nu}$$

which leads to a fractal dimension $D = 3$ for the inside of these surfaces. This would put them into another universality class than the surfaces with a restricted number of handles.

Unfortunately, a Monte Carlo study of SAS with an arbitrary number of handles seems to be in disagreement with these predictions [280]. This study was performed using the techniques for simulating surfaces introduced by O'Connell and coworkers. Since the genus can now be arbitrary, some extra moves should be allowed in the algorithm. Again the exponents of branched polymer behaviour are found: $\nu = 0.506 \pm 0.005$ and $\theta = 1.50 \pm 0.06$. Also, the ratio $\langle V_N \rangle / N$ reaches a constant implying that even in this case $y_{\Delta p} = 1/\nu \approx 2$. The explanation for why the RG argument fails has been traced by Stella *et al.* [280] to the fact that the amplitude of the singularity having dimension $y_{\Delta p} = 3$ is zero. The next leading term in the RG calculation is then precisely that associated with the field K which has a dimension $1/\nu$. In this way one obtains a crossover exponent which is 1, in agreement with the numerical results.

Before finishing this discussion, we make some remarks. Firstly, we want to point out that, as was the case in $d = 2$, one can transform the vesicle model to a model for the collapse of branched polymers in $d = 3$. For a discussion of this and for some numerical results on a vesicle model with a handle fugacity we refer the reader to [280]. Secondly, Stella *et al.* [280] determined the dependence of the exponent θ on the number of handles. For $H = 1$ they found $\theta = -0.5 \pm 0.5$ and for $H = 2$, $\theta = -2.3 \pm 1.0$. While no simple formula is known for the dependence of θ on H, the trend seen here is similar to that for the θ-exponent of c-animals, and to that for the dependence of the exponent α of SAPs on the number of prime knots in the polygon. Finally, we mention that it has been shown by Janse van Rensburg and Whittington [265] that the growth constant for surfaces with $H = 1$ is strictly less than that of surfaces with $H = 0$.

11.3 The crumpling transition

The final topic we discuss in this book is the behaviour of surfaces when the effects of bending rigidity are included. This is an
extremely important effect for the modelling of membranes using
SAS. The model is as follows: for each SAS we count the number
of edges along which the plaquettes incident on that edge are at
right angles. We will call such an edge a fold. This number is denoted by l_p. A partition function, suitable to look for a crumpling
transition, is

$$Z_N^{cr}(\beta) = \sum_{l_p} S_N(l_p) \exp(\beta l_p) \qquad (11.9)$$

where β is a (reduced) bending energy, which can either be positive
or negative, and where $S_N(l_p)$ is the number of surfaces of N
plaquettes with l_p folds. As usual we expect

$$Z_N^{cr}(\beta) \sim \mu_{cr}^N(\beta) N^{-\theta}$$

Clearly, for $\beta = 0$ we recover the branched polymer behaviour
of SAS, known from the previous sections. When $\beta \to -\infty$, the
partition function is dominated by flat surfaces. The transition
between the flat and non-flat, or crumpled, regime can in principle
occur smoothly without passing through a critical point. That
is the case for SAWs. For surfaces there are many calculations,
mainly based on field theoretic approaches (such as an extension
to surfaces of the Edwards model), that show that there should
be a real crumpling phase transition [281]. A lot of effort has
been devoded to finding this transition in models for polymerised
membranes (without much succes). The crumpling transition was
also not found in Monte Carlo studies of fluid membranes in the
continuum [282]. In this section we investigate the situation for
SAS and discuss both rigorous and numerical work. The surprising
result is that in this case, evidence for a crumpling transition has
been found.

Whittington [277] proved some interesting exact results about
the free energy that can be obtained from (11.9)

$$f^{cr}(\beta) = \lim_{N \to \infty} \frac{1}{N} \log(Z_N^{cr}(\beta)) \qquad (11.10)$$

Using concatenation arguments, he showed that this free energy
exists, and is a convex function of β. He also obtained rigor-

ous upper and lower bounds on $\mu_{cr}(\beta) = \exp f^{cr}(\beta)$. An obvious lower bound to (11.9) (for vesicles) is given by the subset of vesicles which are flat 'wafers' of thickness 1. These can easily be mapped onto a two-dimensional vesicle. If each plaquette has a grand canonical fugacity z, the vesicle associated with the wafer has an area fugacity z^2, a step fugacity for the surface $z^2 \exp 2\beta$ and an additional bending rigidity for the SAR which is $\exp \beta$. Now, we know (chapter 9) that the vesicle grand partition function (in absence of bending rigidity) diverges for inflated vesicles. This argument can be extended to the case when the rigidity is included. Therefore when $z^2 > 1$, the free energy of SAS must also diverge. Hence, we obtain a lower bound on the free energy (11.10) of $f^{cr}(\beta) \geq 0$. Moreover, when $\beta \to -\infty$, the free energy is completely determined by the 'wafers'. We can therefore expect that in that limit the critical point of the SAS is related to the line of essential singularities in the $d = 2$ vesicle model. Since at $\beta = 0$ the universality class of the SAS is that of branched polymers (which represents a second transition), we are led to believe that there exists, at some $\beta_c < 0$ a tricritical point which we can interpret as a crumpling transition. The bounds on $f^{cr}(\beta)$ have recently been improved by Janse van Rensburg [283]. This author was able to prove the interesting result that $f^{cr}(\beta) > 0$, which is of some relevance for interpreting the numerical results discussed below.

Numerical work has been performed on closed SAS (or vesicles). Baumgärtner [284] used his canonical Monte Carlo method to study the crumpling model (11.9). He investigated the specific heat of the model which is given by the fluctuation in the total energy of the surfaces. The presence of a peak in the specific heat was then interpreted as evidence for a crumpling transition. From a study of the radius of gyration, or the moments of inertia of the surface, one can try to determine the universality class of the surface at the crumpling transition. This led to a very rough estimate for the exponent $\nu_{cr} = 0.46 \pm 0.025$. Unfortunately, the canonical Monte Carlo method does not allow a determination of the exponent θ.

Most recently an extensive Monte Carlo study of the crumpling transition for closed SAS without handles has been performed by Orlandini and coworkers [285]. The method used was the grand

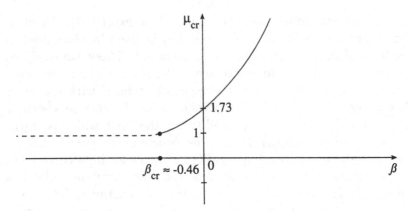

Fig. 11.4. Phase diagram for the crumpling transition. The dotted line indicates a line of first order transitions, while along the full line the transition is second order.

canonical method of O'Connell *et al.*. These authors found convincing evidence for a phase transition at $\beta_{cr} \approx -0.46$. This point can be interpreted as a crumpling transition, since for higher β-values the well known branched polymer exponents were found. At this crumpling transition, the following exponents were estimated; $\nu_{cr} = 0.4825 \pm 0.0015$, $\theta_{cr} = 1.78 \pm 0.03$. These values clearly rule out BP behaviour and seem to be the first clear numerical evidence for SAS in a new universality class. In figure 11.4 we summarise the current knowledge about the behaviour of $\mu^{cr}(\beta)$. While from the numerical work it is impossible to decide whether $\mu^{cr}(\beta \leq \beta_{cr}) = 1$, the lower bound on the free energy mentioned above rules out this possibility. However, the difference between $\mu^{cr}(\beta \leq \beta_{cr})$ and one is so small that it cannot be seen on the scale of figure 11.4.

Finally, we mention that the existence of a crumpling transition in a subset of the set of all SAS can be proven exactly. One defines the skeleton of a surface as the set of all folds. For a general surface this skeleton can be disconnected. Janse van Rensburg [283] recently showed, using exact inequalities and relations with lattice animals, that the set of surfaces with a connected skeleton has a crumpling transition. Moreover, in the flat phase, the free energy is exactly zero.

Further work remains to be done to understand why a crumpling transition occurs for SAS, but seems to be absent in other models for surfaces in the continuum.

References

[1] Eisele U., *Introduction to polymer physics*, Springer-Verlag (1990).

[2] Feynman R., Leighton R., Sands M., *The Feynman Lectures on Physics*, vol. II, 31-1 , Addison-Wesley (1965).

[3] Flory P.J., *Statistical mechanics of chain molecules*, Interscience (1969).

[4] Hughes B.D., *Random walks and random environments*, Clarendon Press (1996).

[5] Claes I., Van den Broeck C., *J. Stat. Phys.* **49**, 383 (1987).

[6] Halpin-Healy T., Zhang Y.C., *Phys. Rep.* **254**, 215 (1995).

[7] Mézard M., Parisi G., Virasoro M.A., *Spin glass theory and beyond*, World Scientific (1989).

[8] Barabási A., Stanley H.E., *Fractal concepts in surface growth*, Cambridge University Press (1995).

[9] Cook J., Derrida B., *J. Stat. Phys.* **57**, 89 (1989).

[10] Derrida B., Golinelli O., *Phys. Rev.* **A41**, 4160 (1990).

[11] Kardar M., *Nucl. Phys.* **B290**, 582 (1987).

[12] Sherrington D., Kirkpatrick S., *Phys. Rev. Lett.* **35**, 1792 (1975).

[13] Bethe H.A., *Z. Phys.* **71**, 205 (1931).

[14] Karabach M., Müller G., *Comp. Phys.* **11**, 36 (1997) and further issues of *Computers in Physics*.

[15] de Gennes P.G., *Scaling concepts in polymer physics*, Cornell University Press (1979).

[16] Hammersley J.M., Morton K.W., *J. Roy. Stat. Soc.* **B16**, 23 (1954).

[17] Madras N., Slade G., *The self avoiding walk*, Birkäuser (1993).

[18] Kesten H., *J. Math. Phys.* **5**, 1128 (1964).

[19] Conway A.R., Guttmann A.J., *Phys. Rev. Lett.* **77**, 5284 (1996).

[20] Enting I.G., Guttmann A.J., *J. Phys.* **A22**, 1371 (1989).

[21] Guttmann A.J., *J. Phys.* **A22**, 2807 (1989).

[22] Guttmann A.J. in *Phase transitions and critical phenomena*, vol.13 edited by C. Domb and J. Lebowitz, Academic Press (1989).

[23] Cardy J., *Scaling and renormalisation in statistical physics*, Cambridge University Press (1996).

[24] Dekeyser R., Maritan A., Stella A.L., *Phys. Rev.* **A36**, 2338 (1987).
[25] Bouchaud J.P., Georges A., *Phys. Rev.* **B39**, 2846 (1989).
[26] Marqusee J.A., Deutch J.M., *J. Chem. Phys.* **75**, 5179 (1981).
[27] Pietronero L., Peliti L., *Phys. Rev. Lett.* **55**, 1479 (1985).
[28] Mandelbrot B.B., *The fractal geometry of nature*, Freeman (1982).
[29] de Gennes P.G., *Phys. Lett.* **A38**, 339 (1972).
[30] Yeomans J.M., *Statistical mechanics of phase transitions*, Oxford University Press (1992).
[31] Stanley H.E., *Phys. Rev.* **176**, 718 (1968).
[32] Nienhuis B., *Phys. Rev. Lett.* **49**, 1062 (1982).
[33] Duplantier B. in *Fundamental problems in statistical mechanics VII*, edited by H. van Beijeren, North-Holland (1990).
[34] Rudin W., *Functional analysis*, McGraw Hill (1973).
[35] Brydges D.C., Spencer T., *Commun. Math. Phys.* **97**, 125 (1985).
[36] Cardy J., Hamber H.W., *Phys. Rev. Lett.* **45**, 499 (1980).
[37] Nienhuis B., *Phys. Rev. Lett.* **49**, 1062 (1982).
[38] Baxter R.J., *J. Phys.* **A19**, 2821 (1986).
[39] Batchelor M.T., Blöte H.W.J., *Phys. Rev. Lett.* **61**, 138 (1988).
[40] Nienhuis B. in *Phase transitions and critical phenomena*, vol. 11, edited by C. Domb and J.L. Lebowitz, Academic Press (1987).
[41] Jackson J.D., *Classical electrodynamics*, Wiley (1975).
[42] Derrida B., *J. Phys.* **A14**, L5 (1981).
[43] Belavin A.A., Polyakov A.M., Zamolodchikov A.B., *J. Stat. Phys.* **34**, 763 (1984).
[44] Cardy J.L., in *Phase transitions and critical phenomena*, vol. 11, edited by C. Domb and J.L. Lebowitz, Academic Press (1987).
[45] Cardy J.L., *J. Phys.* **A17**, L385 (1984).
[46] Trovato A., tesi di laurea, universitá di Padova (1996).
[47] Saleur H., *J. Phys.* **A20**, 455 (1987).
[48] Friedan D., Qiu Z., Shenker S., *Phys. Rev. Lett.* **52**, 1575 (1984).
[49] Blöte H.W.J., Cardy J.L., Nightingale M.P., *Phys. Rev. Lett.* **56**, 742 (1986).
[50] Affleck I., *Phys. Rev. Lett.* **56**, 746 (1986).
[51] Dotsenko Vl. S., Fateev V.A., *Nucl. Phys.* **B240**, 312 (1984).
[52] Cardy J.L., *J. Phys.* **A21**, L797 (1988).
[53] Cardy J.L., Saleur H., *J. Phys.* **A22**, L601 (1989).
[54] Lieb E.H., *Phys. Rev.* **162**, 72 (1967).
[55] Baxter R.J., *Exactly solved models in statistical mechanics*, Academic Press (1982).
[56] Nienhuis B., *Int. J. Mod. Phys.* **B4**, 929 (1990).

212   References

[57] Batchelor M.T., Nienhuis B., Warnaar S.O., *Phys. Rev. Lett.* **62**, 2425 (1989).
[58] Warnaar S.O., Batchelor M.T., Nienhuis B., *J. Phys.* **A25**, 3077 (1992).
[59] Blöte H.W.J. , Nienhuis B., *J. Phys.* **A22**, 1415 (1989).
[60] Batchelor M.T., *J. Phys.* **A26**, 3733 (1993).
[61] Dhar D., *J. Math. Phys.* **19**, 5 (1978).
[62] Elezović S., Knežević M., Milošević S., *J. Phys.* **A20**, 1215 (1987).
[63] Dhar D., *J. Physique* **49**, 397 (1988).
[64] Stanley H.E., Reynolds P.J., Redner S. , Family F. in *Real space renormalisation*, edited by J.M.J. van Leeuwen and T. Burkhardt, Springer-Verlag (1982).
[65] des Cloiseaux J., Jannink G., *Polymers in solution: their modelling and structure*, Oxford University Press (1990).
[66] des Cloiseaux J., *J. Physique* **42**, 635 (1981).
[67] Edwards S.F., *Proc. Phys. Soc.* **85**, 613 (1985).
[68] Le Guillou J.C., Zinn-Justin J., *J. Physique Lett.* **46**, L137 (1985).
[69] Binder K., *Monte Carlo and molecular dynamics simulations in polymer science*, Oxford University Press (1995).
[70] Rosenbluth M.N., Rosenbluth A., *J. Chem. Phys.* **23**, 356 (1955).
[71] Lal M., *Mol. Phys.* **17**, 57 (1969).
[72] Madras N., Sokal A.D., *J. Stat. Phys.* **50**, 109 (1988).
[73] Cox D. R., Miller H. D., *The theory of stochastic processes*, Chapman and Hall (1990).
[74] Li B., Madras N., Sokal A.D., *J. Stat. Phys.* **80**, 661 (1995).
[75] Guim I., Blöte H.W.J., Burkhardt T.W., *J. Phys.* **A30**, 413 (1997).
[76] Madras N., Orlitsky A., Shepp L.A., *J. Stat. Phys.* **58**, 159 (1990).
[77] Berg B., Foerster D., *Phys. Lett.* **B106**, 323 (1981).
[78] Aragão de Carvalho C., Caracciolo S., Fröhlich J., *Nucl. Phys.* **B215**, 209 (1983).
[79] Caracciolo S., Pelissetto A., Sokal A.D., *J. Stat. Phys.* **60**, 1 (1990).
[80] Berretti A., Sokal A.D., *J. Stat. Phys.* **40**, 483 (1985).
[81] Carmesin I., Kremer K., *Macromolecules* **21**, 2819 (1988).
[82] Binder K., in *Phase transitions and critical phenomena*, vol. 8, edited by C. Domb and J.L. Lebowitz, Academic Press (1983).
[83] Burkhardt T.W., Cardy J.L., *J. Phys.* **A20**, L233 (1987).
[84] Barber M.N., *Phys. Rev.* **B8**, 407 (1973).
[85] Eisenriegler E., *Polymers near surfaces*, World Scientific (1993).
[86] De'Bell K., Lookman T., *Rev. Mod. Phys.* **65**, 87 (1993).
[87] Duplantier B., Saleur H., *Phys. Rev. Lett.* **57**, 3179 (1986).

[88] Batchelor M. T., Suzuki J., *J. Phys.* **A26**, L729 (1993).
[89] Cardy J.L., *Nucl. Phys.* **B240**, 514 (1984).
[90] Vanderzande C., *J. Phys.* **A23**, 563 (1990).
[91] Hammersley J.M., Torrie G.M., Whittington S.G., *J. Phys.* **A15**, 539 (1982).
[92] Hegger R., Grassberger P., *J. Phys.* **A27**, 4069 (1994).
[93] Eisenriegler E., *J. Chem. Phys.* **79**, 1052 (1983).
[94] Zhao D., Lookman T., De'Bell K., *Phys. Rev.* **A42**, 4591 (1990).
[95] Burkhardt T.W., Eisenriegler E., Guim I., *Nucl. Phys.* **B316**, 559 (1989).
[96] Meirovitch H., Chang I., *Phys. Rev.* **E48**, 1960 (1993).
[97] Guim I., Burkhardt T.W., *J. Phys.* **A22**, 1131 (1989).
[98] Batchelor M.T., Yung C.M., *Phys. Rev. Lett.* **74**, 2026 (1995).
[99] Bouchaud E., Vannimenus J., *J. Physique* **50**, 2931 (1989).
[100] Kumar S., Singh Y., Dhar D., *J. Phys.* **A26**, 4835 (1993).
[101] Stella A.L., Vanderzande C., *Int. J. Mod. Phys.* **B4**, 1437 (1990).
[102] Stella A.L., Giugliarelli G., *Phys. Rev.* **E53**, 5035 (1996).
[103] Stauffer D., Aharony A., *Introduction to percolation theory*, Taylor and Francis (1992).
[104] Grimmett G., *Percolation*, Springer-Verlag (1991).
[105] Essam J. W., *Rep. Prog. Phys.* **43**, 833 (1980).
[106] Deutscher G., Zallen R., Adler J., *Percolation structures and processes*, Annals of the Israel Physical Society vol. 5, Adam Hilger (1983).
[107] Hammersley J.M., *Annals of Math. Stat.* **28**, 790 (1957).
[108] Peierls R., *Proc. Camb. Phil. Soc.* **32**, 477 (1936).
[109] Kesten H., *Comm. Math. Phys.* **74**, 41 (1980).
[110] Kunz H., Souillard B., *J. Stat. Phys.* **19**, 77 (1978).
[111] Stauffer D., *Phys. Rep.* **54**, 1 (1979).
[112] Havlin S., Bunde A. in *Fractals and disordered systems*, edited by A. Bunde and S. Havlin , Springer-Verlag (1991).
[113] Stanley H. E. in *On growth and form*, edited by H.E. Stanley and N. Ostrowsky, NATO ASI Series, Martinus Nijhoff Publishers (1986).
[114] Saleur H., Duplantier B., *Phys. Rev. Lett.* **58**, 2325 (1987).
[115] Wu F. Y., *Rev. Mod. Phys.* **54**, 235 (1982).
[116] Baxter R.J., *J. Phys.* **C6**, L94 (1973).
[117] Coniglio A., *J. Phys.* **A15**, 3829 (1982).
[118] Harary F., *Graph theory*, Addison- Wesley (1969).
[119] Kasteleyn P. W., *Physica* **29**, 1329 (1963).
[120] Baxter R.J., Kelland S.W., Wu F.Y., *J. Phys.* **A9**, 397 (1976).
[121] den Nijs M., *Phys. Rev.* **B27**, 1674 (1983).

214   References

[122] Duplantier B., Saleur H., *Nucl. Phys.* **B290**, 291 (1987).
[123] Nienhuis B., Riedel E.K., Schick M., *Phys. Rev.* **B27**, 5625 (1983).
[124] Kosterlitz J.M., Thouless D.J., *J. Phys.* **C6**, 1181 (1973).
[125] Duplantier B., *J. Phys.* **A19**, L1009 (1986).
[126] Grassberger P., Hegger R., *Ann. Phys.* (Leipzig) **4**, 230 (1995).
[127] Grassberger P., *J. Phys.* **A26**, 1023 (1993).
[128] Manna S.S., Guttmann A.J., *J. Phys.* **A22**, 3113 (1989).
[129] Barber M.N., *Physica* **48**, 237 (1970).
[130] Duplantier B., David F., *J. Stat. Phys.* **51**, 327 (1988).
[131] Duplantier B., *J. Stat. Phys.* **49**, 411 (1987).
[132] Batchelor M.T., Owczarek A., Seaton K.A., Yung C.M., *J. Phys.* **A28**, 839 (1995).
[133] Blöte H.W.J., Nienhuis B., *Phys. Rev. Lett.* **72**, 1372 (1994).
[134] Batchelor M.T., Suzuki J., Yung C.M. , *Phys. Rev. Lett.* **73**, 2646 (1994).
[135] Kondev J., de Gier J., Nienhuis B., *J. Phys.* **A29**, 6489 (1996).
[136] Batchelor M.T., Blöte H.W.J., Nienhuis B., Yung C.M., *J. Phys.* **A29**, L399 (1996).
[137] Tesi M.C., Janse van Rensburg E.J., Orlandini E., Whittington S.G., *J. Phys.* **A29**, 2451 (1996).
[138] Maritan A., Seno F., Stella A.L., *Physica* **A156**, 679 (1989).
[139] de Gennes P.G., *J. Physique Lett.* **36**, L55 (1975).
[140] Coniglio A., Jan N., Maijd I., Stanley H.E., *Phys. Rev.* **B35**, 3617 (1987).
[141] Duplantier B., Saleur H., *Phys. Rev. Lett.* **59**, 539 (1987).
[142] Seno F., Stella A.L., Vanderzande C., *Phys. Rev. Lett.* **65**, 2897 (1990).
[143] Bradley R.M., *Phys. Rev.* **A41**, 914 (1990).
[144] de Queiroz S.L.M., Yeomans J.M., *J. Phys.* **A24**, L933 (1991).
[145] Prellberg T., Owczarek A., *J. Phys.* **A27**, 1811 (1994).
[146] Privman V., *J. Phys.* **A19**, 3287 (1986).
[147] Ishinabe T., *J. Phys.* **A20**, 6435 (1987).
[148] Maes D., Vanderzande C., *Phys. Rev.* **A41**, 3074 (1990).
[149] Saleur H., *J. Stat. Phys.* **45**, 419 (1986).
[150] Veal A.R., Yeomans J.M., Jug G., *J. Phys.* **A24**, 827 (1991).
[151] Seno F., Stella A.L., *J. Physique* **49**, 739 (1988).
[152] Buldyrev S.V., Sciortino F., *Physica* **A182**, 346 (1992).
[153] Geyer C.J., Thompson E.A., *J. Am. Stat. Assoc.* **90**, 909 (1995).
[154] Tesi M.C., Janse van Rensburg E.J., Orlandini E., Whittington S.G., *J. Stat. Phys.* **82**, 155 (1996).
[155] Stephen J.M., *Phys. Lett.* **A53**, 363 (1975).

[156] Duplantier B., *J. Physique* **43**, 991 (1982).
[157] Duplantier B., *Europhys. Lett.* **1**, 491 (1986).
[158] Duplantier B., *J. Chem. Phys.* **86**, 4233 (1987).
[159] Grassberger P., Hegger R., *J. Chem. Phys.* **102**, 6881 (1995).
[160] Vrbová T., Whittington S.G., *J. Phys.* **A29**, 6253 (1996).
[161] Seno F., Stella A.L., Vanderzande C., *Phys. Rev. Lett.* **61**, 1520 (1988).
[162] Vanderzande C., Stella A.L., Seno F., *Phys. Rev. Lett.* **67**, 2757 (1991).
[163] Vanderzande C., *J. Phys.* **A21**, 833 (1988).
[164] Seno F., Stella A.L., *Europhys. Lett.* **7**, 605 (1988).
[165] Foster D., Orlandini E., Tesi M.C., *J. Phys.* **A25**, L1211 (1992).
[166] Owczarek A.L., Prellberg T., Brak R., *Phys. Rev. Lett.* **70**, 951 (1993).
[167] Bennett-Wood D., Brak R., Guttmann A.J., Owczarek A.L., *J. Phys.* **A27**, L1 (1994).
[168] Cardy J.L., *Nucl. Phys.* **B419**,411 (1994).
[169] Miller J., *J. Stat. Phys.* **63**, 89 (1991).
[170] Bennett-Wood D., Cardy J.L., Flesia S., Guttmann A.J., Owczarek A.L., *J. Phys.* **A28**, 5143 (1995).
[171] Barkema G.T., Flesia S., *J. Stat. Phys.* **85**, 363 (1996).
[172] Koo W.M., *J. Stat. Phys.* **81**, 561 (1995).
[173] Trovato A., Seno F., *Phys. Rev.* **E56**, 131 (1997).
[174] Pfeuty P., Velasco R.M., de Gennes P.G., *J. Physique* **38**, L5 (1977).
[175] Kantor Y., Kardar M., *Phys. Rev.* **E51**, 1299 (1995).
[176] Kantor Y., Kardar M., *Phys. Rev.* **E52**, 835 (1995).
[177] Levin Y., Barbosa M.C., *Europhys. Lett.* **31**, 513 (1995).
[178] Kantor Y., Kardar M., *Europhys. Lett.* **28**, 169 (1994).
[179] Branden C., Tooze J., *Introduction to protein structure*, Garland (1991).
[180] Chan H.S., Dill K.A., *Physics Today* **46**, 2, 24 (1993).
[181] Soteros C.E., *J. Phys.* **A25**, 3153 (1992).
[182] Soteros C.E., Sumners D.W., Whittington S.G., *Math. Proc. Camb. Phil. Soc.* **111**,75 (1992).
[183] Duplantier B., *Phys. Rev. Lett.* **57**, 941 (1986).
[184] Wilkinson M.K., Gaunt D.S., Lipson J.E.G., Whittington S.G., *J. Phys.* **A19**, L811 (1986).
[185] Zhao D., Lookman T., *Europhys. Lett.* **26**, 339 (1994).
[186] Grassberger P., *J. Phys.* **A27**, L721 (1994).
[187] Lubensky T.C., Isaacson J., *Phys. Rev.* **A20**, 2130 (1979).
[188] Klarner D.A., *Can. J. Math.* **19**, 851 (1967).

[189] Gaunt D.S., Sykes M.F., Torrie G.M., Whittington S.G., *J. Phys.* **A15**, 3209 (1982).

[190] Guttmann A.J., Gaunt D.S., *J. Phys.* **A11**, 949 (1978).

[191] Gaunt D.S., Ruskin H., *J. Phys.* **A11**, 1369 (1978).

[192] Gaunt D.S., *Physica* **A177**, 146 (1991).

[193] Parisi G., Sourlas N., *Phys. Rev. Lett.* **46**, 871 (1981).

[194] Yang C.N., Lee T.D., *Phys. Rev.* **87**, 404 (1952).

[195] Ising E., *Z. Phys.* **31**, 253 (1925).

[196] Cardy J.L., *Phys. Rev. Lett.* **54**, 1354 (1985).

[197] Fisher M.E., *Phys. Rev. Lett.* **40**, 1610 (1978).

[198] Miller J.D., De'Bell K., *J. Physique* **I 3**, 1717 (1993).

[199] de Alcantara Bonfim O.F., Kirkham J. E., McKane A., *J. Phys.* **A13**, L247 (1980).

[200] Derrida B., de Seze L., *J. Physique* **43**, 475 (1982).

[201] Derrida B., Stauffer D., *J. Physique* **46**, 1623 (1985).

[202] Duarte J.A.M.S., *J. Phys.* **A19**, 1979 (1986).

[203] Caracciolo S., Glaus U., *J. Stat. Phys.* **41**, 95 (1985).

[204] Janse van Rensburg E.J., Madras N., *J. Phys.* **A25**, 303 (1992).

[205] Madras N., Soteros C. E., Whittington S.G., *J. Phys.* **A21**, 4617 (1988).

[206] Soteros C. E., Whittington S.G., *J. Phys.* **A21**, 2187 (1988).

[207] Whittington S.G., Torrie G.M., Gaunt D.S., *J. Phys.* **A16**, 1695 (1983).

[208] Leibler S., Singh R.R.P. and Fisher M.E., *Phys. Rev. Lett.* **59**, 1989 (1987).

[209] Fisher M.E., *Physica* **D38**,112 (1989).

[210] Fisher M.E., Guttmann A.J., Whittington S.G., *J. Phys.* **A24**, 3095 (1991).

[211] Enting I., Guttmann A.J., *J. Stat. Phys.* **58**, 475 (1990).

[212] Duplantier B., *Phys. Rev. Lett.* **64**, 493 (1990).

[213] Banavar J., Maritan A., Stella A.L., *Phys. Rev.* **A43**, 5752 (1991).

[214] Camacho C.J. and Fisher M.E., *Phys. Rev.* **A46**, 6300 (1992).

[215] Orlandini E., Seno F., Stella A.L., Tesi M.C., *Phys. Rev. Lett.* **68**, 488 (1992).

[216] Marqusee J.A., Deutch J.M., *J. Chem. Phys.* **75**, 5179 (1981).

[217] Madras N., Soteros C.E., Whittington S.G., Martin J.L., Sykes M.F., Flesia S., Gaunt D.S., *J. Phys.* **A23**, 5327 (1990).

[218] Gaunt D.S., Flesia S., *J. Phys.* **A24**, 3655 (1991).

[219] Stella A.L., *Phys. Rev.* **E50**, 3259 (1994).

[220] Sykes M.F., Wilkinson M.K., *J. Phys.* **A19**, 3407 (1986).

[221] Chang I.S., Shapir Y., *Phys. Rev.* **B38**, 6736 (1988).

[222] Derrida B., Herrmann H.J., *J. Physique* **44**, 1365 (1983).
[223] Vanderzande C., *Phys. Rev. Lett.* **70**, 3595 (1993).
[224] Flesia S., Gaunt D.S., Soteros C.E., Whittington S.G., *J. Phys.* **A26**, L993 (1993).
[225] Flesia S., Gaunt D.S., *J. Phys.* **A25**, 2127 (1992).
[226] Flesia S., Gaunt D.S., Soteros C.E., Whittington S.G., *J. Phys.* **A25**, L1169 (1992).
[227] Coniglio A., *J. Phys.* **A16**, L187 (1983).
[228] Seno F., Vanderzande C., *J. Phys.* **A27**, 5813 (1994).
[229] Stella A.L., Vanderzande C., *Phys. Rev. Lett.* **62**, 1067 (1989).
[230] Madras N., Janse van Rensburg E.J., *J. Stat. Phys.* **86**, 1 (1997).
[231] De'Bell K., Lookman T., Zhao D., *Phys. Rev.* **A44**, 1390 (1991).
[232] Janssen H.K., Lyssy A., *Phys. Rev.* **E50**, 3784 (1994).
[233] de Queiroz S.L.A., *J. Phys.* **A28**, 6315 (1995).
[234] Lam P.M., Binder K., *J. Phys.* **A21**, L405 (1988).
[235] Edwards S.F., *Proc. Phys. Soc.* **91**, 513 (1967).
[236] Grosberg A., Nechaev S., in *Advances in polymer science* **106**, 1 (1993).
[237] Alexander J.W., *Trans. Amer. Math. Soc.* **30**, 275 (1928).
[238] Murasugi K., *Knot theory and its applications*, Birkhäuser (1996).
[239] Adams C.C., *The knot book*, Freeman (1994).
[240] Thistlewaithe M.B., in *Aspects of Topology*, edited by I. James and E. Kronheimer, Cambridge University Press (1985).
[241] Ernst C., Sumners D.W., *Proc. Camb. Phil. Soc.* **102**, 303 (1987).
[242] Welsh D., *Colloq. Math. Soc. Janos Bolyai* **60**, 713 (1991).
[243] Janse van Rensburg E.J., Sumners D.W., Wasserman E., Whittington S.G., *J. Phys.* **A25**, 6557 (1992).
[244] Diao Y., *J. knot theory and its ramifications* **2**, 413 (1993).
[245] Delbruck M., *Proc. Symp. Appl. Math.* **14**, 55 (1962).
[246] Sumners D. W., Whittington S.G., *J. Phys.* **A21**, 1689 (1988).
[247] Pippenger N., *Disc. Appl. Math.* **25**, 273 (1989).
[248] Kesten H., *J. Math. Phys.* **4**, 960 (1963).
[249] Janse van Rensburg E.J., Whittington S.G., *J. Phys.* **A23**, 3573 (1990).
[250] Janse van Rensburg E.J., Whittington S.G., *J. Phys.* **A24**, 5553 (1991).
[251] Vanderzande C., *J. Phys.* **A28**, 3681 (1995).
[252] Quake S. R., *Phys. Rev. Lett.* **73**, 3317 (1994).
[253] Janse van Rensburg E.J., Whittington S.G., *J. Phys.* **A24**, 3935 (1991).
[254] Orlandini E., Tesi M.C., Janse van Rensburg E.J., Whittington S.G., *J. Phys.* **A29**, L299 (1996).

[255] Orlandini E., Tesi M.C., Whittington S.G., Sumners D.W., Janse van Rensburg E.J., *J. Phys.* **A27**, L333 (1994).

[256] Sumners D. W., *Math. Intell.* **12**, 71 (1990).

[257] Lacher R. C., Sumners D.W. in *Computer simulations of polymers*, edited by R.J. Row, Prentice-Hall (1991).

[258] Janse van Rensburg E.J., Orlandini E., Sumners D.W., Tesi M.C., Whittington S.G., *J. Phys.* **A26**, L981 (1993).

[259] Kholodenko A.L., Vilgis T. A., *J. Phys.* **A29**, 939 (1996).

[260] Grosberg A. Y., Khokhlov A. R., *Statistical physics of macromolecules*, AIP Press (1994).

[261] Nelson D., Piran T., Weinberg S., *Statistical mechanics of membranes and surfaces*, World Scientific (1989).

[262] Kantor Y., Kardar M., Nelson D., *Phys. Rev. Lett.* **57**, 791 (1986).

[263] Durhuus B., Fröhlich J., Jonsson T., *Nucl. Phys.* **B225**, 185 (1993).

[264] David F. in *Statistical mechanics of membranes and surfaces*, edited by D. Nelson., T. Piran and S. Weinberg, World Scientific (1989).

[265] Janse van Rensburg E.J., Whittington S.G., *J. Phys.* **A22**, 4939 (1989).

[266] Maritan A., Stella A.L., *Nucl. Phys.* **B280**, 561 (1987).

[267] Kogut J., *Rev. Mod. Phys.* **51**, 659 (1979).

[268] Maritan A., Stella A.L., *Phys. Rev. Lett.* **53**, 123 (1984).

[269] Redner S., *J. Phys.* **A18**, L723 (1985).

[270] Sterling T., Greensite J., *Phys. Lett.* **B121**, 345 (1983).

[271] Glaus U., *J. Stat. Phys.* **50**, 1141 (1988).

[272] O'Connell J., Sullivan F., Libes D., Orlandini E., Tesi M.C., Stella A.L., Einstein T. L., *J. Phys.* **A24**, 4619 (1991).

[273] Baumgärtner A., Romero A., *Physica* **A187**, 243 (1992).

[274] Orlandini E., Stella A.L., Tesi M.C., Sullivan F., *Phys. Rev.* **E48**, R4203 (1993).

[275] Maritan A., Seno F., Stella A.L., *Phys. Rev.* **B44**, 2834 (1991).

[276] Cambier J.L., Nauenberg M., *Phys. Rev.* **B34**, 8071 (1986).

[277] Whittington S.G., *J. Math. Chem.* **14**, 103 (1993).

[278] Banavar J., Maritan A., Stella A.L., *Science* **252**, 825 (1991).

[279] Kadanoff L., *Rev. Mod. Phys.* **49**, 267 (1977).

[280] Stella A.L., Orlandini E., Beichl I., Sullivan F., Tesi M.C., Einstein T., *Phys. Rev. Lett.* **69**, 3650 (1992).

[281] Nelson D. in *Statistical mechanics of membranes and surfaces*, edited by D. Nelson, T. Piran and S. Weinberg, World Scientific (1989).

[282] Gompper G., Kroll D.M., *Phys. Rev.* **E51**, 514 (1995).
[283] Janse van Rensburg E.J., *J. Stat. Phys.* **88**, 177 (1997).
[284] Baumgärtner A., *Physica* **A192**, 550 (1993).
[285] Orlandini E., Stella A.L., Einstein T., Tesi M.C., Beichl I., Sullivan F., *Phys. Rev.* **E53**, 5800 (1996).

Index

Printed in the United States
By Bookmasters